Lecture Notes in Electrical Engineering

Volume 987

The book series *Lecture Notes in Electrical Engineering* (LNEE) publishes the latest developments in Electrical Engineering—quickly, informally and in high quality. While original research reported in proceedings and monographs has traditionally formed the core of LNEE, we also encourage authors to submit books devoted to supporting student education and professional training in the various fields and applications areas of electrical engineering. The series cover classical and emerging topics concerning:

- Communication Engineering, Information Theory and Networks
- Electronics Engineering and Microelectronics
- Signal, Image and Speech Processing
- Wireless and Mobile Communication
- Circuits and Systems
- Energy Systems, Power Electronics and Electrical Machines
- Electro-optical Engineering
- Instrumentation Engineering
- Avionics Engineering
- Control Systems
- Internet-of-Things and Cybersecurity
- Biomedical Devices, MEMS and NEMS

For general information about this book series, comments or suggestions, please contact leontina.dicecco@springer.com.

To submit a proposal or request further information, please contact the Publishing Editor in your country:

China

Jasmine Dou, Editor (jasmine.dou@springer.com)

India, Japan, Rest of Asia

Swati Meherishi, Editorial Director (Swati.Meherishi@springer.com)

Southeast Asia, Australia, New Zealand

Ramesh Nath Premnath, Editor (ramesh.premnath@springernature.com)

USA, Canada

Michael Luby, Senior Editor (michael.luby@springer.com)

All other Countries

Leontina Di Cecco, Senior Editor (leontina.dicecco@springer.com)

**** This series is indexed by EI Compendex and Scopus databases. ****

Daljeet Singh · Raghavendra Kumar Chaudhary ·
Krishna Dev Kumar
Editors

Computer Aided Constellation Management and Communication Satellites

Proceedings of the International Conference on Small Satellites, ICSS 2022

Springer

Editors
Daljeet Singh
Department of Research and Innovation
Division of Research and Development
Lovely Professional University
Phagwara, Punjab, India

Raghavendra Kumar Chaudhary
Department of Electrical Engineering
IIT Kanpur
Kanpur, Uttar Pradesh, India

Krishna Dev Kumar
Department of Aerospace Engineering
Ryerson University
Toronto, ON, Canada

ISSN 1876-1100 ISSN 1876-1119 (electronic)
Lecture Notes in Electrical Engineering
ISBN 978-981-19-8554-6 ISBN 978-981-19-8555-3 (eBook)
https://doi.org/10.1007/978-981-19-8555-3

This Springer imprint is published by the registered company Springer Nature Singapore Pte Ltd.
The registered company address is: 152 Beach Road, #21-01/04 Gateway East, Singapore 189721, Singapore

Preface

This book Vol-III entitled *Computer Aided Constellation Management and Communication Satellites*, presents select proceedings of International Conference on Small Satellites (ICSS 2022) and aims to provide an opportunity for academicians, scientists, industry experts, and researchers engaged in teaching, research, and development to present and discuss ideas and share their views on communication satellites and computer aided constellation management. This book offers a broad-spectrum picture of different aspects of communication satellites including intersatellite communication, personal communication, audio and video broadcasting, Internet through satellites, TT&C link design, and many more. The thematic book issue also presents the emerging research on constellation management of more than 2,500 active satellites orbiting the earth. Computer aided constellation management will enable information delivery and knowledge exchange without the historical delays associated with terrestrial relays, human-in-the-loop decision making, and ground-based processing. The applications for these small satellite missions range from asset tracking, earth observation, defense-related space programs, communications, space science, environment monitoring, and other applications of small satellites. The edited thematic volume book will be a valuable asset for beginners, researchers, and professionals working in the field of communication satellites and computer aided constellation management.

Phagwara, India — Daljeet Singh
Kanpur, India — Raghavendra Kumar Chaudhary
Toronto, Canada — Krishna Dev Kumar

Contents

About the Editors

Dr. Daljeet Singh is working as Assistant Professor in the Centre for Space Research, Division of Research and Development at Lovely Professional University, India. He received the B.Tech. (Hons.) and M.Tech. in Electronics and Communication Engineering from Lovely Professional University, India, in 2011 and 2013, respectively, and the Ph.D. degree in Electronics and Communication Engineering from Thapar Institute of Engineering and Technology, India, in 2019. His current research interests include the design, optimization and fabrication of planar microstrip antennas for UWB, satellite and other communication applications. He is also working on the mathematical modeling, characterization and simulation of wireless communication systems including MIMO-OFDM, massive MIMO, NOMA, spatial multiplexing and simulation of advanced wireless communication systems by utilizing MATLAB and Simulink platforms. He has guided three M.Tech. and two Ph.D. students, and currently, several M.Tech./Ph.D. students are working under him. He was awarded Institutional Fellowship during his Ph.D. and M.Tech. He has a good track record of publications in top-tier conferences and journals like *IEEE*, Elsevier and Wiley, *AEU*, *PIERS*. He has published six research patents and holds one design patent. He is an active reviewer of various international journals of repute.

Dr. Raghavendra Kumar Chaudhary is an Associate Professor of the Department of Electrical Engineering, Indian Institute of Technology (IIT) Kanpur, India. He served in the Department of Electronics Engineering, IIT Dhanbad, as Assistant Professor (June 2013–April 2021) and as Associate Professor (April 2021–May 2022). He received a B.Tech. from UIET Kanpur, India, in 2007, an M.Tech. from IIT (BHU) Varanasi, India, in 2009, and a Ph.D. from IIT Kanpur, India, in 2014. He has researched developing the metamaterial antenna and dielectric resonator antenna (DRA). The development of the circularly polarized (CP) compact antenna is one of his major areas of contribution. He has published over 250 papers in leading journals and conferences along with a reference book on CP-DRA published by Artech House, London, UK. He has guided 19 M.Tech. students and 11 Ph.D. students, and currently, several M.Tech./Ph.D. students are working under him. He is a Recipient of the Young Scientist Platinum Jubilee Award (2021) of the National Academy of

Sciences, India (NASI), the Young Engineers Award (2020) of the Indian National Academy of Engineering (INAE), the Young Scientist Award (2020) of the Institution of Electronics and Telecommunication Engineers (IETE), Young Engineers Award (2019–2020) of the Institution of Engineers, India (IEI), and many best paper awards in different categories in national and international conferences. He has served as Student Chair for the IEEE Student Branch of Uttar Pradesh Section, in 2012–2013, as Advisor of the IEEE Student Branch, and as Founding Advisor of the AP/MTT Societies Student Branch Joint Chapter of IIT (ISM) Dhanbad. He was also a Member of the Executive Committee of IEEE Kolkata Section, 2021, and an Executive Committee Member of IEEE AP-MTT Kolkata Chapter, 2022. He is serving as Associate Editor of three journals, namely *IET Microwave Antennas and Propagation, IEEE Access and Microwave and Optical Technology Letters*, Wiley. He is a Senior Member of IEEE, Senior Member of URSI, INAE Young Associate and a Life Member of InRaSS. He is a Potential Reviewer of several international journals and also got recognition from IEEE Antenna and Propagation Society for his exceptional performance as a Reviewer of IEEE Transactions on Antenna and Propagation for 2020–2021. He has also been featured and interviewed by IET Electronics Letters, UK.

Dr. Krishna Dev Kumar received his Ph.D. degree in aerospace engineering from the Indian Institute of Technology, Kanpur, India, in 1998. He is a Professor of Aerospace Engineering and Director of the Artificial Intelligence for Aerospace Systems (AIAS) Laboratory at Ryerson University, Canada. His research interests are in the areas of spacecraft dynamics and control, fault diagnosis and prognosis, big data, predictive analytics and artificial intelligence. He has received several awards that include Member of the International Academy of Astronautics, France (2019), Eminent Alumnus Award, Veer Surendra Sai University of Technology, Sambalpur (2017), Sarwan Sahota Ryerson Distinguished Scholar Award (2015), Canada Research Chair (2005–2015), Associate Fellow and Life Member of AIAA (2012), Ontario Early Researcher Award (2006–2011), Japan Society for the Promotion of Science (JSPS) Fellowship (2001–2003) and Science and Technology Agency (STA) Fellowship (1998–2000). He has more than 230 research publications in national and international journals and conferences.

A Microstrip Patch Antenna Using SIR Technique Designed for C-Band Satellites Application

Rupali, Sanjay Kumar Sahu, and Gopinath Palai

Abstract The development of compact, light, high-efficiency, low-cost antennas for wireless communication systems, as well as their integration with the rest of the system, has become a tempting task. Microstrip patch antennas are a popular, low-cost design. The demand for more durable and compact patch structures is increasing as a result of satellite communication technology services. The suggested antenna displayed a return loss response of less than −10 dB starting at 6.5 GHz with a VSWR < 2. Reduced ground plane provided a good tuning parameter for return loss. The available antenna designs work at frequencies ranging from 4 to 8 GHz, making them appropriate for C-band satellite communications. This C-band satellite antenna is primarily used in military applications for speech and data communication. The design and evaluation processes were replicated using HFSS software.

Keywords Microstrip · 4–8 GHz · C-band · Military satellite · Satellite communication · HFSS · SIR

1 Introduction

Nowadays, in satellite communication, there is a need for the creation of antennas having features like ultra-compact, low cost, low profile, high gain, direction, and so on. Antennas connect transmitting and receiving equipment and are used as a path for space propagation. While general principles may apply to satellite communications antennas, enforced gain, field pattern, and, most importantly, physical environment constraints result in design requirements that must be considered. The main

Rupali (✉) · S. K. Sahu
Lovely Professional University, Phagwara, India
e-mail: rupalipuri92@gmail.com

S. K. Sahu
e-mail: sanjay.23393@lpu.co.in

G. Palai
Gandhi Institute for Technological Advancement, Bhubaneswar, India
e-mail: gpalai28@gmail.com

D. Singh et al. (eds.), *Computer Aided Constellation Management and Communication Satellites*, Lecture Notes in Electrical Engineering 987,
https://doi.org/10.1007/978-981-19-8555-3_1

purpose of designing a satellite antenna in such a way is to provide coverage over a defined restricted area, defining antenna gain constraints along with small size and lightweight. Due to these qualities, planar antennas, in particular, may find considerable use on communications satellites. Since the late 1970s, the international antenna community has invested significant effort in theoretical and practical research on microstrip and printed antennas that benefit from low profile, compatibility with integrated circuit technology, and conformability to a curved surface. Antennas can be used on aircraft and rockets, as well as in commercial applications like mobile satellite communications and the Direct Broadcast Satellite (DBS) system. They can also be used for things like GPS and remote sensing, but their main use is on ground transceivers.

The last two decades have been particularly beneficial for research on planar antennas, mostly for mobile communications, while its relevance to satellite systems should be expanded. Various papers based on microstrip patch antennas which are applicable for C-band space communication systems [1]. In this paper, a circular microstrip patch antenna having a concentric diamond-shaped slot is used to improve the parameters like bandwidth and gain compared to a conventional circular microstrip patch antenna, along with fulfilling all requirements for antennas used in satellite communication systems. In addition [2], in this paper, a circular patch antenna with fractals has been fed with an L probe feeding technique that produces a dual-band application for C-band application. Apart from this, in the last decades, various scholars have worked on different performance parameters like return loss, bandwidth, etc. that are useful in Wi-Fi, radar, and military applications.

Numerous antennas operate successfully in the low-frequency band due to their resonant nature. It is critical to tune the antenna to the same band as the RF system to which it is attached in order to prevent transmission and reception from being impaired. So, this article discusses the performance of small microstrip patch antennas using "Arlon AD255C (tm)" in experiments. The antenna operates optimally between 4 and 8 GHz and can also be utilized in the C-band for communication signaling. The intended paper is now divided into six additional sections. Sections 2–4 describe the brief introduction of techniques used to design purposed antenna for satellite communications systems, Sects. 5 and 6 design and the simulation findings, and Sect. 7 summarizes the research.

2 Microstrip Patch Antenna

A microstrip patch antenna (MPA) is a dielectric substrate with one side containing a conducting patch of any planar or non-planar geometry and the other including a ground plane. It is a frequently used printed resonant antenna for microwave wireless communications in the narrow band that requires semi-hemispheric coverage. Due to its planar configuration and simplicity of integration with microstrip technology, the microstrip patch antenna has been extensively explored and is frequently used as an element in an array. Numerous microstrip patch antennas have already been investigated. A thorough collection of geometries is provided, along with their unique properties. Rectangular and circular patches are the most basic and frequently used microstrip antennas. These patches are designed for the simplest and most demanding applications. Rectangular geometries are naturally separable and easy to analyze. The circular patch antenna has the benefit of emitting radiation in a symmetrical pattern [3].

2.1 Antenna Parameter Formula

The equation can be used to get the width (w) of the microstrip patch antenna and the patch's effective dielectric constant (Fig. 1).

3 Fractal Antenna

The most recent study on the use of fractal geometries in antenna design demonstrates that these antennas have low sidelobe levels. The popularity of these antennas is due to their electrically vast structure, which conveniently packs into small spaces [4]. Fractal geometries are generated due to the fractal structure's self-similarity and the symmetrical character of the structure surrounding a point [4]. Several fractal designs, such as Sierpinski Carpet [5], Sierpinski Gasket [6, 7], Koch Loop [8], and Hilbert Curve, have been brought into the field of antennas in recent years to improve antenna properties. Several of the geometries have been effective in terms of antenna size reduction, while others are aimed at multi-band response (see Fig. 2). For optimal performance, a size reduction of 2–4 times is feasible. The primary objective of fractal antenna engineering is to push the boundaries of Euclidean geometry in terms of antenna design and synthesis [4].

$$Width = \frac{c}{2f_o\sqrt{\frac{\varepsilon_R+1}{2}}};\quad \varepsilon_{eff} = \frac{\varepsilon_R+1}{2} + \frac{\varepsilon_R-1}{2}\left[\frac{1}{\sqrt{1+12\left(\frac{h}{W}\right)}}\right]$$

$$Length = \frac{c}{2f_o\sqrt{\varepsilon_{eff}}} - 0.824h\left(\frac{(\varepsilon_{eff}+0.3)\left(\frac{W}{h}+0.264\right)}{(\varepsilon_{eff}-0.258)\left(\frac{W}{h}+0.8\right)}\right)$$

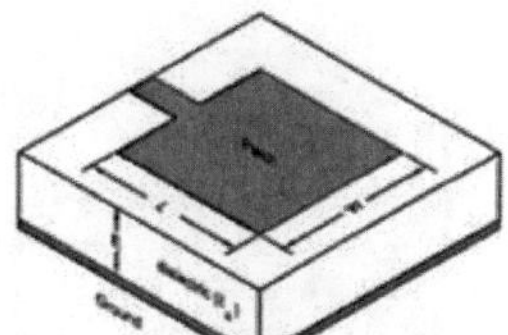

Fig. 1 Antenna parameter formula

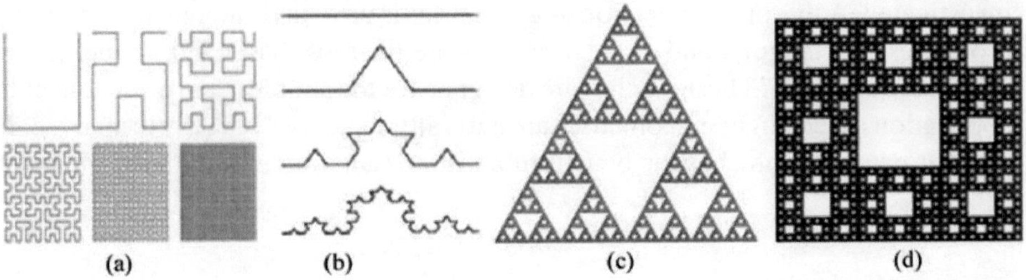

Fig. 2 Types of fractal shapes; **a** Hilbert curve, **b** Koch loop, **c** Sierpinski gasket [9], **d** Sierpinski carpet

4 Stepped-Impedance Resonator

A SIR is composed of numerous sections with varying lengths and alternating impedances that alter the transmission line's impedance and direct current to favored locations along its length. SIR is a technique used in the design of microwave components that enables component downsizing [10]. The fundamental structures of SIR are depicted in Fig. 3.

Where Z_1 denotes the impedance of a narrow portion, Z_2 denotes the impedance of a wider segment, and θ_1 and θ_2 respectively denote the electric lengths of narrow and broad sections.

$$\frac{Z_2}{Z_1} = \tan \beta_1 l_1 \tan \beta_2 l_2 = K \quad (1)$$

where θ_1 denotes the electrical length of the section with characteristic impedance Z_1, and the formula is as follows:

$$\theta_1 = \beta_1 l_1 \quad (2)$$

And θ_T denotes the transmission line's total electrical length, which is given by

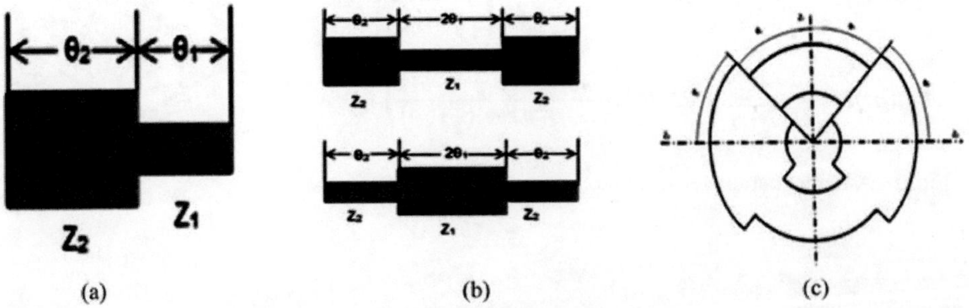

Fig. 3 Basic structures of SIR [11, 12]; **a** quarter wavelength ($\lambda g/4$) type **b** half wavelength ($\lambda g/2$) type **c** one wavelength (λg) type

Table 1 Dimensions of antenna design

S. No.	Parameters	Values (mm)
1	a_1	2.275
2	a_2	0.525
3	a_3	1.4875
4	a_4	6.825

$$\theta_T = \theta_1 + \theta_2 \quad (3)$$

θ_T is minimum when

$$\theta_1 = \theta_2 = \tan^{-1}\sqrt{K} \quad (4)$$

$$\theta_{T(\min)} = \tan^{-1}\left(\frac{2\sqrt{K}}{1-K}\right) \quad (5)$$

From (5), we obtain the resonator's minimal line length for $K < 1$.

5 Antenna Design

Thus, this unit-cell structure is modeled using the "High-Frequency Structure Simulator (HFSS)," a three-dimensional electromagnetic structure simulator, with the help of a fractal spiral resonator whose structure has been modified via the SIR approach. This is a simulator based on the finite element method (FEM), which is used to determine the electromagnetic behavior of a design and provides information on numerous modeling and simulation elements. By spiraling together, the fractal spiral resonator is created. Rings are mirrored versions of the first order Hilbert fractal. Through inductive and capacitive coupling between the various components, the resonance frequency must be lowered. The design is applied to an "Arlon AD255C (tm)" substrate with a relative permittivity of 2.55, a thickness of 1.524 mm, and dimensions of 8.75 * 8.75 mm. The length of each line's broader portion in SI-FSR is calculated by multiplying the line's length by 0.3216, its placement is centered along its corresponding line, and the dimensions of SI-FSR are given in Table 1 (Fig. 4).

6 Result and Discussion

It has been designed to operate in the C-band that is used for satellite communication. The suggested antenna's performance parameters, including as return loss, VSWR,

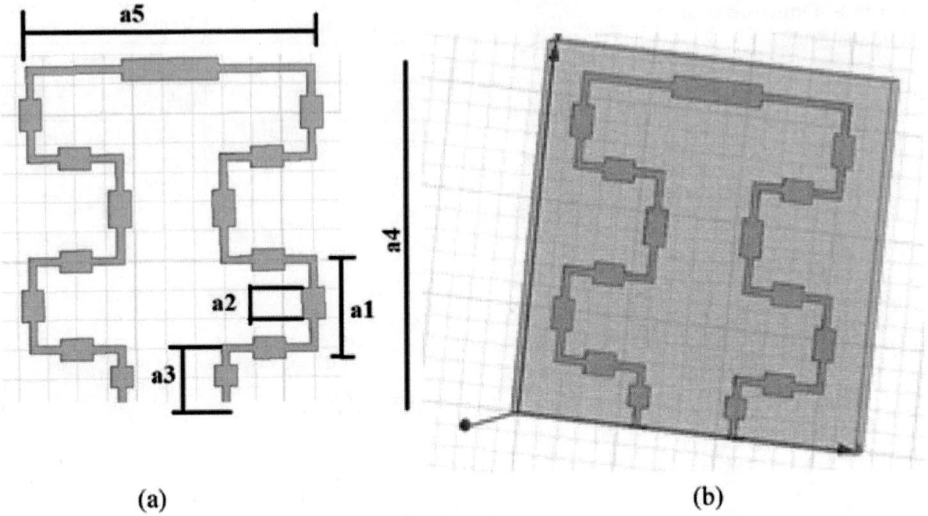

Fig. 4 **a** Geometry of SI-FSR, **b** SI-FSR geometry encased in radiation box

and gain, have been simulated using HFSS and demonstrate good agreement with the threshold values for each parameter.

6.1 Return Loss

Figure 5 depicts the simulation result for the planned military-based antenna's returned loss signal. The suggested microstrip patch antenna's operating frequency is 6.5 GHz, as determined through simulation.

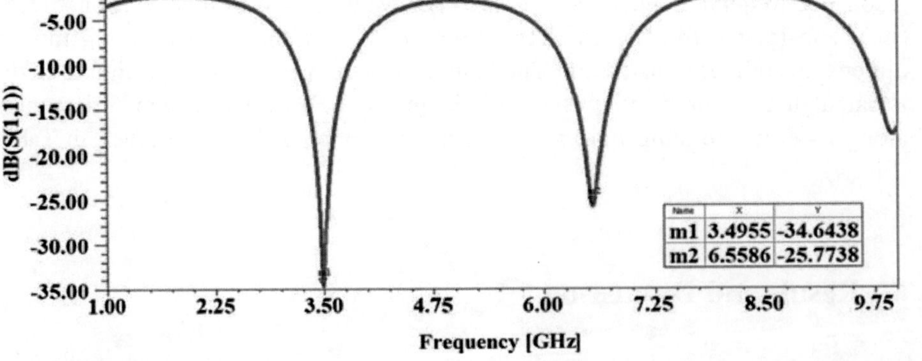

Fig. 5 Reflection coefficient (S_11) of SI-FSR

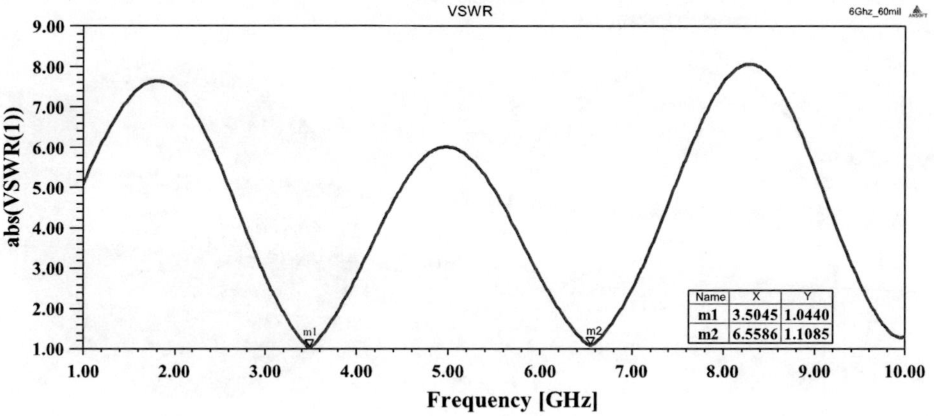

Fig. 6 Voltage standing wave ratio of SI-FSR

6.2 Voltage Signal Wave Ratio (VSWR)

Figure 6 illustrates the VSWR of the C-band microstrip antenna. The band's simulated VSWR is less than 2.

6.3 Radiation Pattern

Figure 7 depicts the simulation result for three-dimensional radiation patterns and antenna parameters of SI-FSR.

7 Conclusion

The purpose of this work is to design and simulate a C-band antenna. The research proposes a signal loss and radiation pattern for their returned signal in the far field. According to the HFSS software results, a return loss in C-band patch antenna is above 10 dB. As a result, this design is capable of meeting a variety of criteria for satellite-based military communication applications. The VSWR as well as the improvement factor for signal return loss will be explored currently in order to achieve high gain and small size with increased packing efficiency onboard military satellite systems. The future scope will detail the procedure for increasing the width of such an antenna.

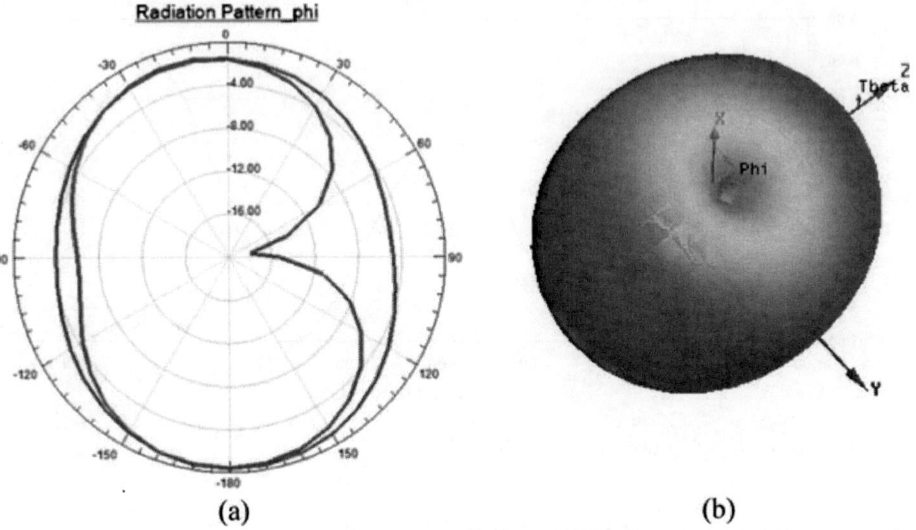

Fig. 7 **a** Two-dimensional radiation pattern of SI-FSR, **b** three-dimensional radiation pattern of SI-FSR

References

1. Matin MA, Sharif BS, Tsimenidis CC (2007) Probe fed stacked patch antenna for wideband applications. IEEE Trans Ant Propag 55(8)
2. Bhatnagar D, Saini JS, Saxena VK, Joshi LM (2011) Design of circular patch antenna with diamond shape slot. Indian J Radio Space Phys
3. Balanis CA (1997) Antenna theory analysis and design. John Wiley & Sons, Inc
4. Mandelbrot BB (1983) The fractal geometry of nature. W.H. Freeman Company
5. Naghshvarian-Jahromiv (2008) Novel wideband planar fractal monopole antenna. IEEE Trans Ant Propag 56(12):3844–3849
6. Borja Borau C, Puente Baliarda C, Median MA (1998) Iteration network model to predict the behaviour of a Sierpinski fractal network. IEEE Electron Lett 34:1443–1445
7. Cohen N (1996) Fractal antennas: part 1, 2. Commun Q (Summer: 7(22):53–66)
8. Puente Baliarda C, Romeu Robert J, Pous Andrés R, Ramis J, Hijazo A (1998) A small but long Koch fractal monopole. Electron Lett 34(9)
9. Puente-Baliarda C, Romeu J, Pous R, Cardama A (1998) On the behaviour of the sierpinski multi band fractal antenna. IEEE Trans Ant Propag 46(4):517–524
10. Makimoto M, Yamashita S (2001) Microwave resonators and filters for wireless communication—theory, design and application. Springer-Verlag, Berlin, Heidelberg
11. Makimoto M, Yamashita S (1979) Compact band pass filters using stepped impedance resonators. Proc IEEE 67(1):16–19
12. Agawa M, Makimoto M, Yamashita S (1997) Geometrical structures and fundamental characteristics of microwave stepped-impedance resonators. IEEE Trans Microw Theory Tech 45(7)

Split Ring Resonator-based Conformal Antenna for Earth Coverage Spaceborne Applications

G. Shine Let, M. Nesasudha, N. M. Sivamangai, and S. Sridevi Sathya Priya

Abstract This paper describes the design of a split ring resonator-based conformal antenna for Earth coverage spaceborne applications. To have a reduction in antenna size without degradation in the performance, two design techniques are included in the antenna design. First, a rectangular slot is etched from the circular radiating patch. Secondly, with a partial ground plane in the antenna design, a circular split ring resonator has been engraved. The substrate used for the projected antenna design is a flexible Rogers substrate (RO3006) having a relative permittivity of 6.15 and a thickness of 1.27 mm. The overall antenna size is 20 mm × 15 mm × 1.27 mm. Without conformal structure, the designed antenna provided a wide frequency of operation from 4.73 to 6.12 GHz, a peak gain of 1.9 dBi, and a radiation efficiency of 99.8 percent. The patch design is wrapped over a cylinder having a 10 mm radius along the y-axis to obtain the bending structure. The designed conformal antenna has an operating frequency range from 4.54 to 5.91 GHz. For Earth coverage applications and aircraft applications, the designed antenna can be used.

Keywords Split ring resonator · Conformal antenna · Spaceborne applications · Circular radiating patch · Rectangular slot

G. S. Let (✉) · M. Nesasudha · N. M. Sivamangai · S. S. S. Priya
School of Engineering and Technology, Karunya Institute of Technology and Sciences, Coimbatore, India
e-mail: shinelet@gmail.com

M. Nesasudha
e-mail: nesasudha@karunya.edu

N. M. Sivamangai
e-mail: sivamangai@karunya.edu

S. S. S. Priya
e-mail: sridevi@karunya.edu

D. Singh et al. (eds.), *Computer Aided Constellation Management and Communication Satellites*, Lecture Notes in Electrical Engineering 987,
https://doi.org/10.1007/978-981-19-8555-3_2

1 Introduction

Flexible antennas play a major role in flexible electronics, aircraft, medical telemetry, body-worn applications, and military, remote sensing applications [1]. The antenna is designed should be wrapped along the curved surface of the aircraft or communication modules. Different materials such as flexible silicon rubber [2], polydimethylsiloxane (PDMS) [3, 4], polytetrafluoroethylene [5], Rogers Duroid 5880LZ [6], paper, and textile materials are used as flexible dielectric substrates. Based on the applications, different flexible materials are chosen as substrates [7]. Copper, gold, and polymers are used as flexible conducting materials. By comparing with the non-conformal structure, the deviation in conformal antenna performance such as radiation characteristics, gain, and reflection coefficient should be very less. Various flexible radiating structures are used to design the antenna for the required operating frequency. In [8], a rectangular loop shape conformal antenna is designed in Roger's 5870 substrate and is inserted inside a capsule. This antenna design performance is tested for wireless endoscopy applications.

A circular wheel-shaped single-negative metamaterial is used in the antenna design for wireless applications [9]. Using unit-cell analysis, the type of metamaterial is finalized. For 5G Wi-Fi applications, polyethylene terephthalate (PET)-based flexible substrate with highly conductive graphene is used for the design of the antenna in [10]. In [11], a resonator-type antenna is fabricated on a flexible Rogers 3850HT substrate, and the analysis is carried out for on-body applications. Flexible antennas used for wearable or on-body applications have used meander structures. Using different meander structures, compactness in the antenna is achieved.

For Earth coverage application, eight PIFA antenna elements are placed in a circular ring to generate a circularly polarized isoflux beam [12]. A circular shape coaxial feed conformal patch shape design is proposed in [13]. This antenna operates at 1.189 and 1.575 GHz which are used for global navigation satellite systems. To analyze the antenna performance, the designed antenna is wrapped on the top of the aircraft and tested. For spaceborne applications, stair-shaped dual-feed waveguide structures are proposed to achieve high gain [14]. In [15], to have the antenna performance in the spaceborne application, H-shaped and C-shaped slots are incorporated into the antenna design. In [16], four helical U-shape strips are connected with a cuboid shape polyimide substrate. This design works in the ultra-high frequency range. In literature, various shapes and materials are used for the antenna design based on the applications. In our proposed work, a flexible Rogers dielectric substrate is used due to the advantages such as low dielectric loss, impedance matching enhancement, and the low release of vapor/gas for spaceborne applications. The details of the proposed work are explained in detail in the following sections.

2 Design of Split Ring Resonator-Based Conformal Antenna

The proposed split ring resonator (SRR)-based conformal antenna structure is shown in Fig. 1. A circular shape radiating structure with a rectangular slot is proposed in the design. Microstrip feed is given to the circular radiating patch through a transmission line of size AF × BF mm^2. The radiating patch and ground plane design are carried out in a flexible copper conducting sheet of a thickness of 0.035 mm. The dielectric substrate used in the design is flexible Rogers substrate (RO3006) which has a relative permittivity of 6.15 and a thickness of 1.27 mm. The dimension of the split ring resonator-based conformal antenna is 20 × 15 mm^2. A partial ground plane structure is proposed.

A dual parasitic circular split ring resonator is incorporated into the design to achieve the required frequency of operation. By changing the position of the rectangular slot in the circular patch, the current distribution is altered. Using parametric study, the position of the slot in the circular patch is incorporated in the design. The slot is placed below 2.3 mm from the top of the circular patch. The radius of the circular patch is 6.3 mm. The rectangular slot size in the radiating patch is AS × BS mm^2. Using the combination of a rectangular slot and SRR structures, compactness in the design is achieved. The dual parasitic circular SRR structure is kept at a distance of 3 mm from the partial ground plane. The two SRRs are kept at a distance of 0.5 mm. The split size in the SRR is 2 mm × 1 mm. The other dimensions used in the design are given in Table. 1. Using parametric studies, SRR-based conformal antenna dimensions are obtained. The results achieved by the proposed antenna are discussed in the next section.

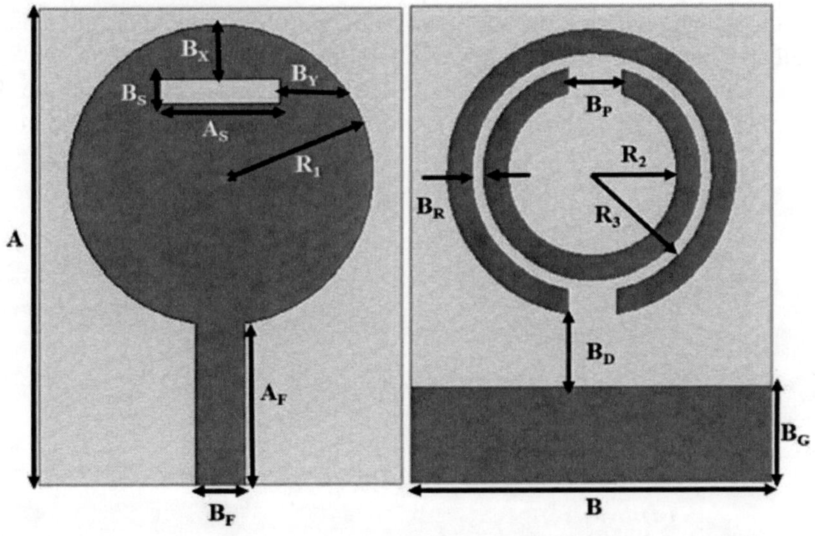

Fig. 1 SRR-based conformal antenna structure **a** radiating structure **b** ground plane

Table 1 Design parameters of SRR-based conformal antenna

Parameter	**A**	**B**	$\mathbf{B_F}$	$\mathbf{A_F}$	$\mathbf{A_S}$	$\mathbf{B_S}$	$\mathbf{B_X}$	$\mathbf{B_Y}$
Dimensions (mm)	20	15	2	6.8	5	1	2.3	2.5
Parameter	$\mathbf{R_1}$	$\mathbf{R_2}$	$\mathbf{R_3}$	$\mathbf{B_R}$	$\mathbf{B_D}$	$\mathbf{B_G}$	$\mathbf{B_P}$	
Dimensions (mm)	6.3	3.5	5	0.5	3	4	2	

3 Results and Discussions

The proposed SRR-based conformal antenna is designed in a flexible Rogers substrate. Figure 2 shows the proposed antenna trimetric structure without bending and with bending. For getting the curvature structure, the antenna is wrapped in a cylinder of radius 10 mm along the *y*-axis. The property of the cylinder used for wrapping the proposed antenna is the vacuum. So the use of a cylinder does not have any impact on the antenna performance. Without bending, the antenna operates in the frequency range from 4.73 to 6.12 GHz by considering the reflection coefficient characteristics (S_{11}) at −10 dB. By providing a bending radius of 10 mm, the antenna operating range is from 4.54 to 5.91 GHz. There is a 3.38% deviation of higher operating frequency concerning without bending and with bending. At the lower operating frequency, 4 percent deviation without and with bending. Figure 3 depicts the SRR-based conformal antenna S_{11} characteristics with and without bending.

The current distribution of the proposed SRR-based conformal antenna is depicted in Fig. 4. The maximum current density of the proposed antenna is 118.9 A/m. The current distribution in an antenna purely depends on the given electric field. The current distribution and radiation characteristics of the antenna are analyzed at a frequency of 5.5 GHz. This resonant frequency is also used for WLAN applications. Using the rectangular slot, the current distribution of the antenna is altered and with a small size, the required frequency of operation is achieved.

The radiation characteristics of the SRR-based conformal antenna are depicted in Fig. 5. The pattern of the antenna is plotted in the *XZ* plane and *XY* plane. Better

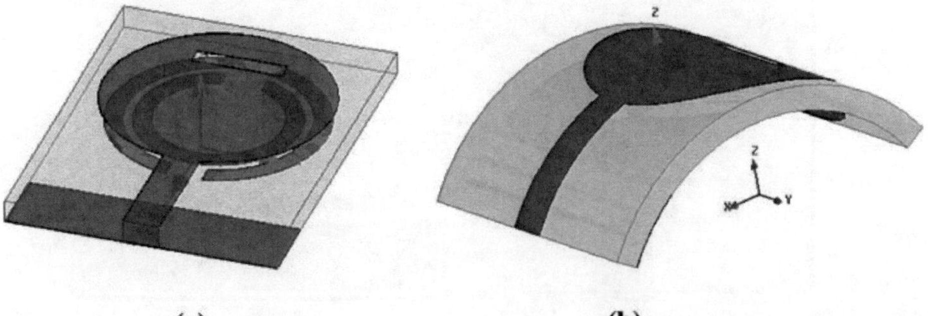

(a) **(b)**

Fig. 2 SRR-based conformal antenna structure **a** without bending **b** with bending

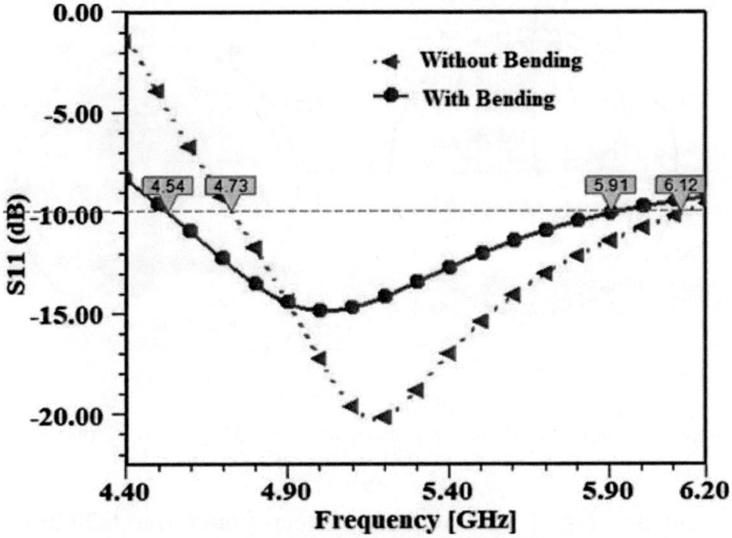

Fig. 3 SRR-based conformal antenna S_{11} characteristics with and without bending

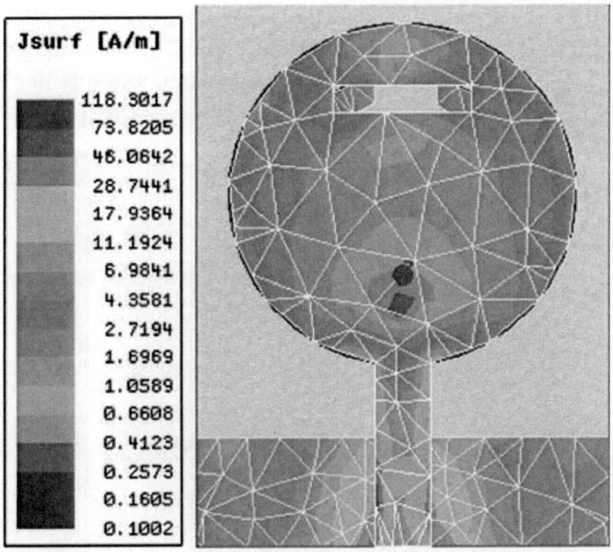

Fig. 4 Current distribution–SRR-based conformal antenna

dipole far-field radiation characteristics are achieved in both planes. In the front view and top view, a figure of 8 shape is obtained. No fields are present at the center of the antenna and no side lobes are present in the obtained radiation characteristics.

The total 3D gain characteristics of the antenna SRR-based conformal antenna is analyzed and depicted in Fig. 6. The antenna's radiation efficiency and peak

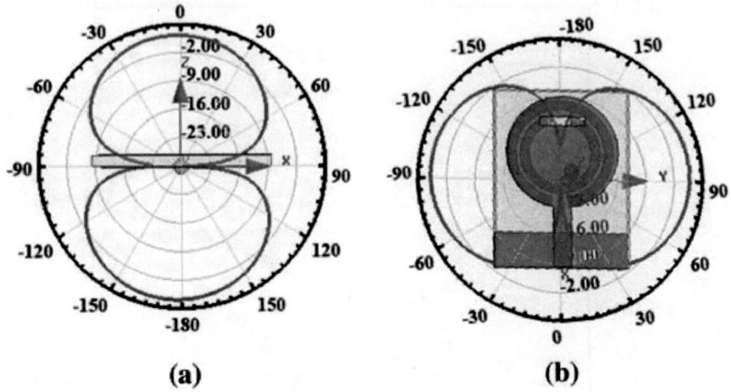

Fig. 5 Radiation pattern–SRR-based conformal antenna **a** *XZ* plane **b** *YX* plane

gain are plotted in Fig. 7 for a frequency sweep from 4.6 to 6.2 GHz. The recommended antenna's radiation efficiency is between 96 and 100% over the whole sweep frequency range. Since the efficiency is high, power loss is very minimal for the designed antenna. Also, there is an absence of back radiating lobes and side lobes due to high radiation efficiency. The peak gain of the proposed antenna is between 1.5 and 2 dB for the frequency sweep of 4.6–6.2 GHz. The antenna gain is proportional to the antenna's directivity. As a result, the suggested antenna has good directivity.

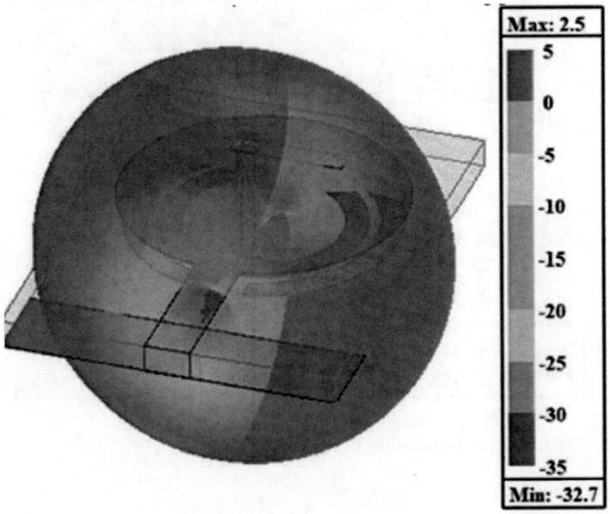

Fig. 6 3D gain pattern-SRR-based conformal antenna

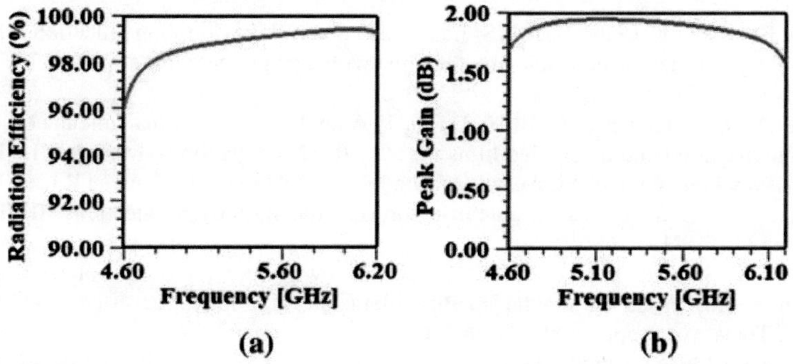

Fig. 7 SRR-based conformal antenna **a** radiation efficiency in % **b** peak gain (dB)

4 Conclusion

In this paper, an SRR-based conformal antenna is proposed. The antenna operates without bending from 4.73 to 6.12 GHz, and with bending from 4.54 to 5.91 GHz. A rectangular slot in the radiating patch and a circular SRR structure in the ground plane contributes to the antenna's compactness. The antenna is designed in flexible Rogers substrate (RO3006) having a thickness of 1.27 mm and the antenna size is 20 mm × 15 mm. The radiation efficiency is greater than 98% for the entire operating frequency range. Antenna gain is between 1.5 and 2 dB for the operating frequency range. The antenna current and radiation characteristics are analyzed for the resonant frequency of 5.5 GHz. Based on the obtained antenna performance, the proposed SRR-based conformal antenna can be used for Earth coverage spaceborne applications.

References

1. Li MJ, Li M, Liu YF, Geng XY, Li YY (2022) A review on the development of spaceborne membrane antennas. Space: Sci Technol:17–20
2. George NM, Pushpa TA, Mary J (2022) Durable silicon rubber-based miniaturized antenna with concentric circle structure for a medical telemetry application. Progr Electromagn Res M 107:155–165
3. Zerith AT, Nesasudha M (2020) A compact wearable 2.45 GHz antenna for WBAN applications. In: IEEE 5th international conference on devices, circuits and systems (ICDCS), pp 184–187
4. Neebha TM, Nesasudha M, Janapala DK (2020) A stable miniaturised AMC loaded flexible monopole antenna for ingestible applications. Comp Biol Med 116
5. Kim S, Shin H (2019) An ultra-wideband conformal meandered loop antenna for wireless capsule endoscopy. J Electromagn Eng Sci 19(2):101–106
6. Balderas LI, Reyna A, Panduro MA, Del Rio C, Gutiérrez AR (2019) Low-profile conformal UWB antenna for UAV applications. IEEE Acc 7:127486–127494
7. Monne MA, Lan X, Chen MY (2018) Material selection and fabrication processes for flexible conformal antennas. Int J Ant Propag
8. Shang J, Yu Y (2020) An ultrawideband and conformal antenna for wireless capsule endoscopy. Microw Opt Technol Lett 62(2):860–865

9. Yang FM, Peng L, Liao X, Mo KS, Jiang X, Li SM (2019) Coupling reduction for a wideband circularly polarized conformal array antenna with a single-negative structure. IEEE Ant Wirel Propag Lett 18(5):991–995
10. Hu Z, Xiao Z, Jiang S, Song R, He D (2021) A dual-band conformal antenna based on highly conductive graphene-assembled films for 5G WLAN applications. Materials 14(17)
11. Nikolayev D, Joseph W, Skrivervik A, Zhadobov M, Martens L, Sauleau R (2019) Dielectric-loaded conformal microstrip antennas for versatile in-body applications. IEEE Ant Wirel Propag Lett 18(12):2686–2690
12. Li S, Liao S, Yang Y, Che W, Xue Q (2021) Low-profile circularly polarized isoflux beam antenna array based on annular aperture elements for CubeSat earth coverage applications. IEEE Trans Ant Propag 69(9):5489–5502
13. Yinusa KA (2018) A dual-band conformal antenna for GNSS applications in small cylindrical structures. IEEE Ant Wirel Propag Lett 17(6):1056–1059
14. Ren X, Wong H (2019) A differentially fed dual-polarized antenna for satellite applications. In: IEEE 8th Asia-Pacific conference on antennas and propagation (APCAP), pp 460–461
15. Fang X, Wang W, Huang GL, Luo Q, Zhang H (2019) A wideband low-profile all-metal cavity slot antenna with filtering performance for space-borne SAR applications. IEEE Ant Wirel Propag Lett 18(6):1278–1282
16. Xue K, Liao S, Zhu R, Xue Q, Ding L, Wang Y, Guo Q (2021) Spaceborne miniaturized UHF dual band helix antenna with a small frequency ratio. Microw Opt Technol Lett 63(6):1767–1773

Antenna Deployment Mechanism for a 3U CubeSat Project

S. Sushir, K. Ullas, Komal Prasad, and Vipul V. Kumar

Abstract There is an ever-increasing requirement for smaller and more efficient satellites. Minimizing the size of satellites has its advantages. It equally imposes several challenges in bringing up the satellite from the drawing board to a flight model. This paper focuses on an antenna deployment mechanism that is being developed by a student satellite team from BMS College of Engineering. Antenna deployment mechanisms of similar class satellite programs were referred and a model has been developed and analyzed. Experiments were also conducted on a prototype model to verify the conceptual working of the mechanism.

Keywords 3U CubeSat · Antenna deployment mechanism · Experiments on antenna deployment

1 Introduction

BMSCE Upagraha is a 3U student satellite project currently under development by the students of BMS College of Engineering. The BMSCE Upagraha has an RGB imaging camera as its payload, and for the communication of the satellite to the ground station, it makes use of two dipole antennas. The antenna plays an integral role in establishing proper communication between the satellite and the ground station. Due to restrictions in the fairing volume, antennas are stowed in the spacecraft body during launch and are deployed in space. Hence, deployment of the antenna adhering to its functional requirements is a crucial phase in the satellite operation. This paper focuses on the development of the antenna deployment mechanism. A conceptual model is designed according to the design criteria established. A prototype model has been fabricated to verify the working of the deployment mechanism. Also mentioned in this paper are the description of the selected system, design considerations, structural analysis, and experiments carried out to verify the concept. Details of activities to be carried out are also given in this paper.

S. Sushir (✉) · K. Ullas · K. Prasad · V. V. Kumar
B.M.S College of Engineering, Basavanagudi, Bengaluru 560019, India
e-mail: sushir.me18@bmsce.ac.in; upagraha@bmsce.ac.in

D. Singh et al. (eds.), *Computer Aided Constellation Management and Communication Satellites*, Lecture Notes in Electrical Engineering 987,
https://doi.org/10.1007/978-981-19-8555-3_3

2 Functional Requirements

It is very much necessary for the mechanism to fulfill the basic requirements throughout the flight period.

- The basic requirement is to keep the antenna in stowed condition during launch and deploy them in orbit.
- 2 dipole antennas are to be deployed, they are the ultra high frequency (UHF) and very high frequency (VHF) antennas. Their specifications are as follows
- For Up-link Receiver (VHF):
  Operating frequency = 145 MHz
  Length (L) = $0.48 * \lambda = 1$ m
  where λ is the wavelength of the transmission
- For Downlink Receiver (UHF):
  Operating frequency = 437 MHz
  Length (L) = $0.48 * \lambda = 0.33$ m
- Angle of deployment of each antenna should be 90°.
- Launch vehicle considered is PSLV. The levels identified are as per the PSLV user manual.
- Stowed natural frequency should be greater than 135 Hz, and a deployed natural frequency of 2 Hz (TBD).
- Quasi-static loads requirements are longitudinal load-7 g compression, 3.5 g tension, and lateral load- ± 2 g.
- Provision should be made through telemetry to monitor the hold-down release and deployment function.

3 Options Study

A study was conducted on some of the previously used deployment mechanisms in different satellites. Figure 1 depicts the different satellites and their deployment mechanisms studied for the option studies. Table 1 provides the details of these mechanisms and their disadvantages.

Considering the pros and cons of all the above-mentioned mechanisms, a new mechanism has been developed which abides by the design criteria. This newly developed mechanism is discussed in detail below.

(a) UCAELFIN [4] (b) SAMSAT-218D [3] (c) XATCOBEO [1] (d) BIRDS 2 CUBESAT [2]

Fig. 1 Options study performed

Table 1 Option studies performed [1–4]

Sr. No.	Factors	Tuna-can deployment mechanism	Fusible element mechanism	Polymeric 3D printed antenna deployment mechanism	Heat wire release mechanism
1	Components used for retention	Fishing line	Soldered alloy and gate	Nylon thread	Dyneema/Vectran
2	Length of the antenna that can be deployed	450–500 mm	300–500 mm	300–500 mm	Around 500 mm
3	Configuration or orientation	45° to the top panel	Parallel to the top panel	Parallel to the top panel	Parallel to the top panel
4	Functioning	Requires two main parts, individual antenna holders and the tuna-can. Once command is provided, a single burn resistor cuts the retention wire and antennae deploys simultaneously [4]	No burning mechanism involved Requires a cradle to hold the antenna, uses a heating element and gate. The heating element melts the soldering, hence deploying the antenna [3]	Antennae are coiled on the extremity of the setup. A single burn mechanism releases all the stowed antennae [1]	Antenna is held down around the mounting screws in the stowed position One side of the antenna is screwed to the secure part The free end is tied with a Dyneema/Vectran cable connected to nichrome wire of the burner circuit [2]
5	Heritage	Used in UCAELFIN Satellite [4]	Has been used in SamSat-218D 3U CubeSat [3]	Used in XATCOBEO [1]	Used in BIRDS-2 CUBESAT [2]

(continued)

4 System Description

Based on the options study performed and the functional and identified antenna requirements, a conceptual model was developed and the details are as follows:

- Lever: 25-gauge stainless steel sheet metal
- Antenna: A 32-gauge stainless steel tape measure
- (Dimensions: 500 mm × 12 mm × 0.2 mm)
- Hold-down wire: Nylon thread
- Base plate
- Lever
- Antenna

Table 1 (continued)

Sr. No.	Factors	Tuna-can deployment mechanism	Fusible element mechanism	Polymeric 3D printed antenna deployment mechanism	Heat wire release mechanism
6	Disadvantages	(i) The height of this mechanism is 13.5 mm, and this should be the minimum clearance on the top panel of the spacecraft. (ii) Does not meet the angle of deployment requirement	(i) Heritage of this mechanism shows that this mechanism has failed to deploy (ii) Since it uses a soldered element as a retention mechanism a standard soldering procedure and soldering skill is required while operating and testing	(i) Outgassing and ESD risks of the sub-chassis are more (ii) The materials used for most of the components are out of the ordinary which gives us less information to work with	(i) Overheating of nichrome wire (ii) Would require considerably higher power

- Antenna holder/housing
- Hold-down wire
- Burn resistor or heating element

The base plate holds the complete mechanism for deploying 4 antennas (2 dipole antennas). In the stowed condition as shown in Fig. 2, the lever is closed which holds the antenna in a coiled form, i.e., in the stowed position. The antenna holder/housing helps in giving the circular shape in this stowed condition. One end of the lever is held with a pivoted joint, with the help of an internally threaded fastener and an external smooth surface, which allows the lever to rotate about this joint. The other end is tied to the hold-down wire which is in turn attached to the adjacent lever similarly. A burn resistor is set up in between two adjacent levers.

This burn resistor is placed in such a way that the hold-down wire is in contact with the burn resistor. A command is then executed by the telemetry, tracking, and command team, so that the burn resistor is activated and current passes through it. The temperature of the burn resistor increases, once the temperature of the burn resistor is greater than the melting point of the hold-down wire, the hold-down wire is cut and the internally stored energy of the coiled antenna pushes the lever making it rotate about the pivot joint. Thereby the antenna uncoils and straightens itself. Figure 3b shows the deployment directions of the lever, as the current is passed through the

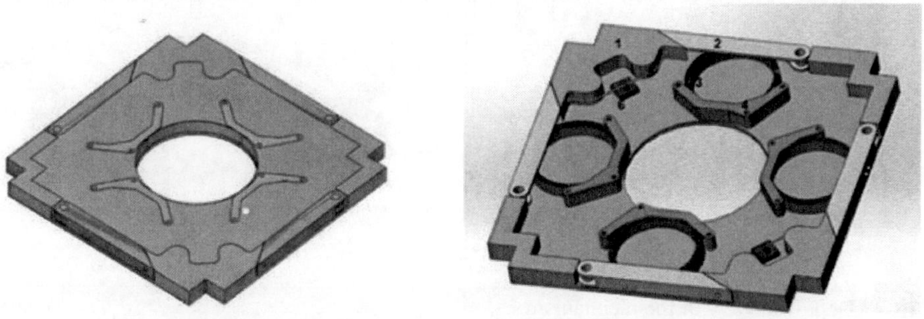

(a) 3D model of Deployment mechanism (b) Detailed view of the deployment mechanism

Fig. 2 Views of the antenna deployment mechanism

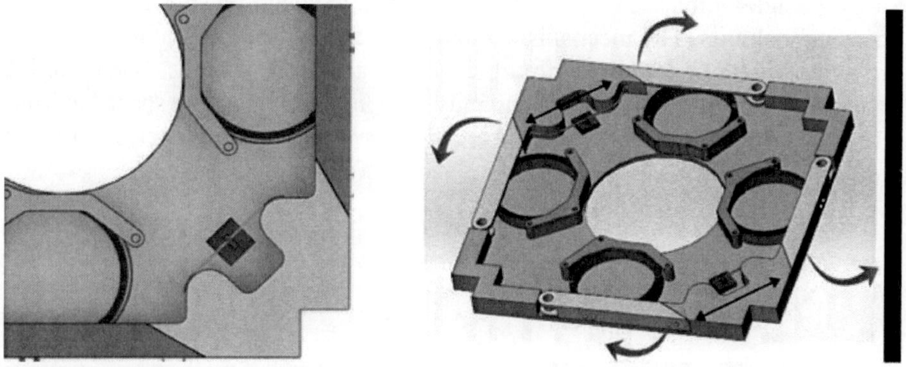

(a) Hold down wire connected to 2 adjacent levers (b) Deployment directions

Fig. 3 Hold-down details and deployment direction

burn resistor, the wire is cut and the internally stored energy of the coiled antenna pushes the lever and the antenna uncoils itself into a stiffened antenna.

Figure 4 represents the deployed state of the antenna, where two long antennae represent 2 very high frequency (VHF) antennas and 2 short antennas represent ultra high frequency (UHF) antennas.

5 Design Considerations

Both the UHF and VHF antennas will be divided into 2 monopoles, each of length of 0.5 m for the VHF band, and 0.17 m for the UHF band. The dimensions of the top panel are $100 * 100\ mm^2$, and a cutout of $8.5 * 8.5\ mm^2$ has to be made on each corner of the satellite's top surface to include the rail rod. The maximum available diameter for the coiling of the antenna in the stowed condition is 25 mm. The angle

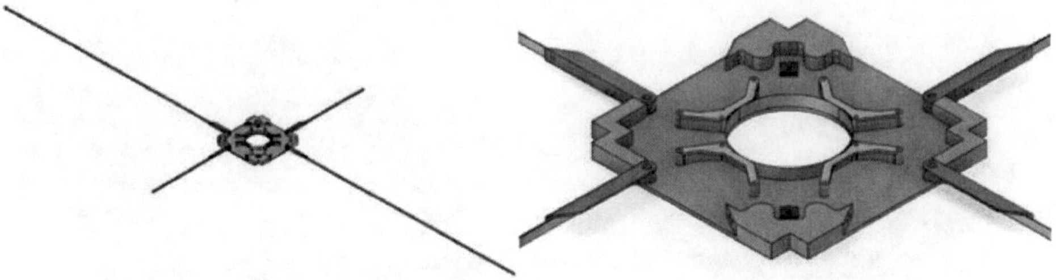

Fig. 4 Deployed view of the mechanism

of deployment of each of the antennas is 90°. Tension in the hold-down wire has to be designed to withstand the launch loads. The deployment process of the mechanism is by a single-point release on command. Deployed latch-up shock should be within acceptable limits. The mechanism should operate in the temperature range of − 30 to + 80 °C and under low earth orbit (LEO) vacuum conditions. The material used must have a good flight heritage and must be compatible with the space environment. The design should meet the stowed and deployed natural frequencies of the antenna and the launch loads as mentioned in the functional requirements.

6 Experiment Conducted

As explained in the system description, after arriving at the basic configuration, a prototype model was realized with available materials. Initial trials were conducted to see the deployment. It was observed that the lever which was manually made out of sheet metal was flexible. This was modified by fabricating it out of a stainless steel block of size 47 × 10 × 8 mm. The hold-down and release mechanism were actuated by manually cutting the wire instead of using the burn wire mechanism.

The parts were assembled on a base plate and deployment trials were conducted. A measuring tape was used to depict the antenna. A coil radius of 25 mm was given to it and the deployment was conducted. Certain observations were made.

It was observed that the antenna during deployment was not deploying smoothly, instead, it used to go back and forth in a whiplash fashion and significant oscillations were observed about the deployment axis. (This phenomenon was found by conducting another test and monitoring the same with the help of a high-speed camera. When the captured data was replayed, it was clearly seen that the deployment was not smooth and the antenna deployed with a whiplash phenomenon.)

To overcome the drawbacks observed in the previous model, a detailed fabrication of the deployment components was done. A high-speed camera was used to capture footage in slow motion. The mass of the lever was slightly increased and the width of the antenna was reduced to avoid whiplash. The new experimental model consists of the following:

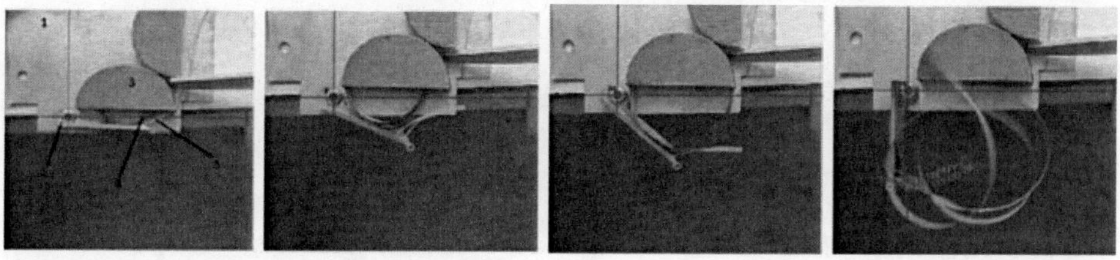

Fig. 5 Tracking of the deployed lever

- Base plate—It was made out of stainless steel material to obtain sufficient rigidity. It houses the other components.
- Lever, which was earlier fabricated by a sheet metal, is now modified to stainless steel. It was fabricated by EDM machining.
- Antenna housing—Acrylonitrile Butadiene Styrene (ABS) material.
- The component was 3D printed.
- The antenna—A 32-gauge stainless steel measuring tape (Dimensions: 500 mm × 4 mm × 0.2 mm) whose width earlier was 12 mm is now reduced to 4 mm.

After monitoring the modification, one more test was conducted and the problems encountered in the earlier test were not seen. The process of the deployment was recorded again using a high-speed camera. Several trials of the deployment were recorded and later analyzed using the 'Tracker' software. It is video analysis and modeling tool. The slow motion video was uploaded into the software and calibrated accurately for every test.

Figure 5 shows the tracking of points of a specified location on the lever with respect to an origin during the deployment.

The origin is set on the axis of rotation of the lever (shown in pink color). The points which are tracked are depicted as cyan diamond markers. The movement of the lever while being deployed is noted and marked at every frame and this path is followed. The software analyzes the movement of the lever, calculating its instantaneous position at every frame with respect to the origin.

A graph of angular velocity (ω) versus time (t) is plotted. Linear curve fitting is done to obtain a best fit straight line for the obtained set of points. By finding the slope of this best fit line, we will have the angular acceleration (a) with which the lever deploys.

$$T = I * a \tag{1}$$

where

T Torque or turning moment experienced by the lever (N-m)
I Moment of inertia about rotating axis of the lever (Kg-m^2)
a Angular acceleration of the lever (rad/s^2).

Slope of the best fit line from Fig. 6 is, (a) = 239 rad/s^2

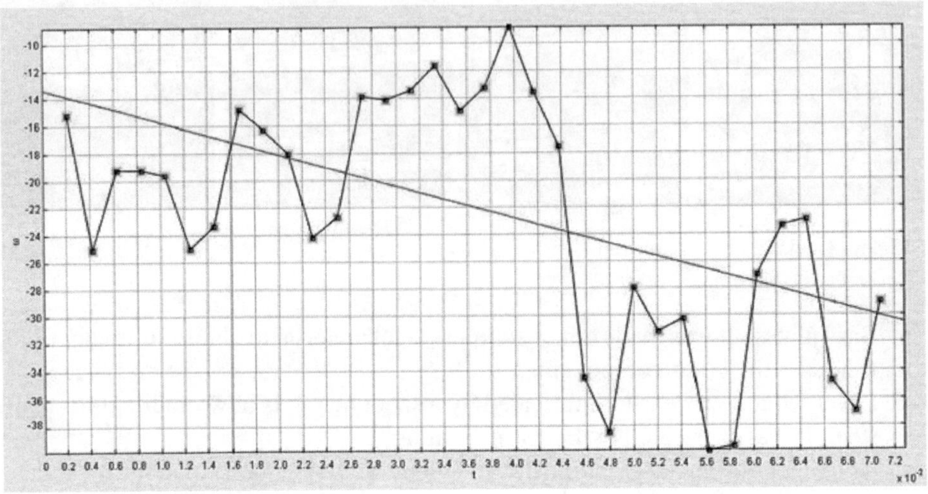

Fig. 6 Graph of ω versus t

The moment of inertia of the lever is found to be $I = 4578.23\ \mathrm{e}^{-9}$ kg-m^2

(This was obtained by finding the mass of the lever and this mass is used as an input to a CAD software from which we obtain the moment of inertia about the rotational axis.)

$$T = I * a = 239 * 4578.23\mathrm{e}^{-9} = 1.094\,\text{N-mm}$$

This value of torque is used as an input in the analysis phase, where a turning moment is given to the internal surface of the lever which would be in contact with the antenna.

7 Analysis Carried Out on the Antenna Deployment Lever in the Stowed Configuration

The types of analysis performed on the lever of the antenna deployment mechanism are:

- Quasi-static acceleration
- Pre-stress modal
- Random vibration.

The load values for all the analysis are taken from the PSLV launch load data. Loads and stiffness for the CubeSat are as follows:

- Longitudinal load: 7 g compression, 3.5 g tension
- Longitudinal stiffness: > 135 Hz

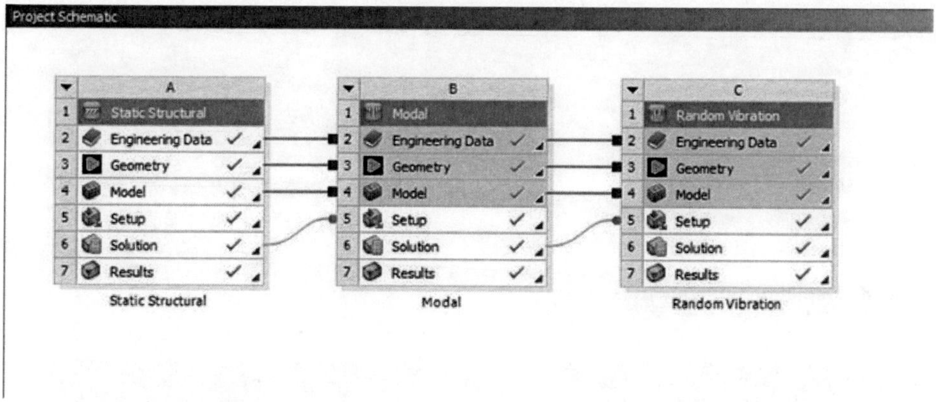

Fig. 7 Analysis description

- Lateral load: ± 2 g
  Lateral stiffness: > 70 Hz
- Qualification factor: 1.25.

This analysis aims to find the stresses developed in the lever, initial modes of vibration, and the reaction forces developed at those places where the hold-down wire will be in contact with the lever (These reaction forces depict the tension to which the hold-down wire will be subjected to.)

The static structural analysis accommodates the quasi-static acceleration loads and this is set as a precursor to the modal analysis, where we get an estimate of the natural frequencies of the setup. This analysis is only helpful in finding the stresses developed on the lever alone and not the antenna as the analysis does not completely qualify the conceptual model as only the lever of the antenna is modeled and the coiled antenna itself is not modeled. The natural frequency of the overall setup is yet to be found (Analysis description is shown in Fig. 7.)

7.1 Geometry

The model for the analysis comprises 3 components (as shown in Fig. 8):

- Base
- Lever
- Through hole fastener.

The lever rotates about the axis of the through hole fastener and is in contact with the external surface of the fastener. The fastener is threaded inside but has a smooth exterior. The base supports both the lever and the fastener. To simplify the geometry and to decrease the computing time, a single lever setup is considered instead of the entire 4 lever assembly. Figure 9 represents the depiction of meshed geometries.

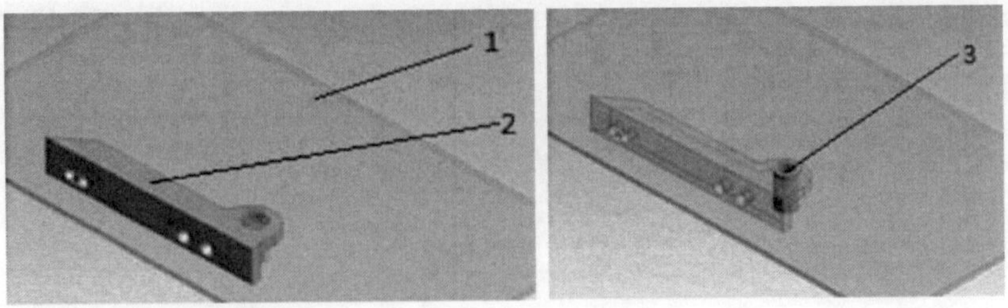

Fig. 8 Depiction of geometries

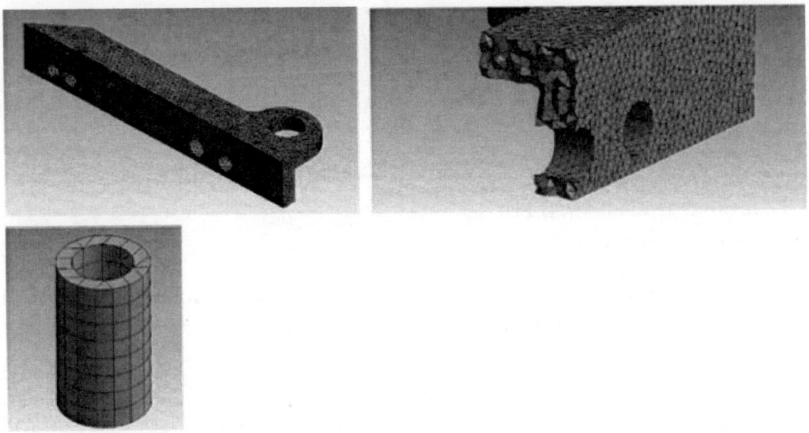

Fig. 9 Depiction of meshed geometries

7.2 Boundary Conditions

- Fixed support: The surface of the lever which will be in contact with the hold-down wire is given a fixed support.
- Moment: A moment of 0.2 N-m is given to the entire surface of the lever which will be in contact with the antenna.
- Remote displacement: A remote displacement is given to that surface which will be in contact with the through hole fastener. To simulate the deployment direction, linear movement of the lever in *X*, *Y*, and *Z* direction are constrained and it is allowed to rotate about the axes.
- Fixed support 2: The threaded or the inner surface of the fastener is given a fixed support since this is the area that will be completely fixed and will not move with respect to the base. Figure 10 depicts the different boundary conditions applied to the meshed geometries.

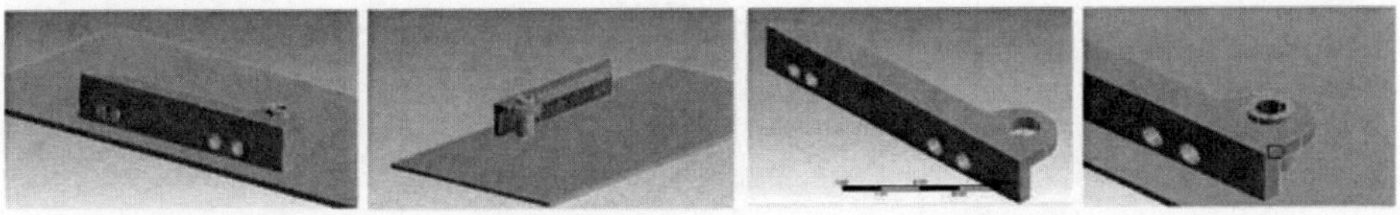

Fig. 10 Boundary conditions

7.3 *Results*

Quasi-static acceleration

Equivalent stresses (Von Mises stress)

Maximum stress of 2.79 MPa is observed at the edges where the hold-down wire is tied to the lever. This depicts that the location where the hold-down wire is attached to the lever is the location of maximum stress concentration under acceleration loads as shown in Fig. 11a.

Reaction force

The total reaction force observed at this support is 6.09 N, i.e., during the application of acceleration loads, the tension in the hold-down wire due to this lever would be around 6.09 N as depicted in Fig. 11b.

Pre-stressed modal results

The first 6 modes of natural frequencies of the lever-fastener-base assembly are presented in Table 2.

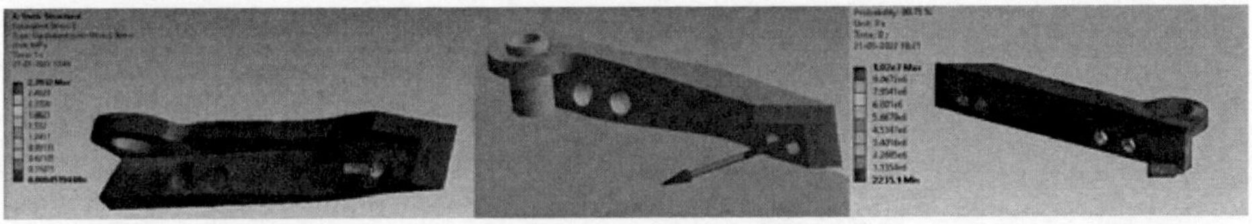

Fig. 11 Depiction of results

Table 2 First 6 modes of vibration

Mode	Frequency (Hz)
1	1211.9
2	2553.9
3	4277.2
4	16,462
5	19,834
6	21,688

Random vibration

- Equivalent stresses (Von Mises stress): Maximum stress of 10.2 MPa is observed at the bottom face edge of the lever. This may be because this face of the lever will be in contact with the base and due to excitation of the vibrational loads, the lever will strike the base plate. This is depicted in Fig. 11c.
- Reaction force: A maximum reaction force of 8.27 N is observed at the hold-down support of the lever faces.

Summary of the analysis performed

- Stresses developed due to the acceleration loads being within acceptable limits.
- The stiffness requirements are met.
- Stresses developed due to random vibration loads are within acceptable limits.
- The reaction forces observed due to the acceleration and random vibration loads would be used for the design of the hold-down system.
- The design of the lever has been decided to be modified so that the bottom face of the lever would not be in contact with the base plate. This is done to reduce the frictional forces in play during deployment.

8 Planned Activities

Detailed design of the hold-down and release mechanism, as well as the deployment mechanism, is to be performed. To provide sufficient stiffness to the antenna, the cross section of the antenna has to be suitably designed (preferably by providing a curvature) to meet the deployed stiffness requirements. Detailed analysis to establish the stowed and deployed natural frequencies and any latch-up shock arising after deployment. Fabrication of the different parts is planned to be carried out using CNC machines. After fabricating the antenna, it has to be suitably heat treated. Beryllium copper was found to be a suitable candidate material for this purpose. After the prototype model is completed, an engineering model will be developed followed by a flight model. Functional tests will be carried out to demonstrate the capabilities of the hold-down release and deployment mechanism. The developed deployment mechanism would be integrated into the satellite structure along with the other components. The assembled satellite structure would be subjected to different environmental tests such as the Vibration test, Shock test, and Thermovac test.

9 Conclusion

The BMSCE Upagraha, which is a 3U satellite, makes use of 2 dipole UHF and VHF antennas. These antennas are in stowed condition during launch and later they are deployed in space by ground command. The mechanism being developed will have a hold-down and release mechanism to keep the antenna in stowed condition during launch and a deployment mechanism to positively release the antenna in orbit.

Briefed in this paper are activities carried out for the development of a hold-down release and antenna deployment mechanism. This includes the option studies performed on existing antenna deployment mechanisms, description of the selected and designed system, design considerations, experiments, and the analysis carried out so far, and also includes the plan for future activities. The results obtained from the initial structural analysis qualify the designed deployment mechanism. Further, the results obtained from this analysis will be used for the detailed design of the mechanism. The option studies performed and the results obtained in this paper will hence be a good platform for the development of the intended deployment mechanism and also for other emerging strategies in this field.

Acknowledgements The authors and the team of students working on this project would like to thank our mentor Shri. C. D. Sridhar and the expert's panel for their constant guidance and support. We also like to extend our thanks to Dr. H. N. Suma, Dr. Jayanthi K. Murthy, the deputy project directors, and BMS College of Engineering for providing this project opportunity and for their continuous encouragement and support.

References

1. Vilán Vilán JA, Aguado Agelet F, Lopez Estevez M, González Muiño A (2012) Antenna deployment mechanism for the Cubesat Xatcobeo. Lessons, evolution and final design. In: Aerospace mechanisms symposium 2012
2. Zaki SBM, Azami MH, BIRDS-2 Project Members, Yamauchi1 T, Kim S, Masui H, Cho M (2018) Design, analysis, and testing of monopole antenna deployment mechanism for BIRDS-2 CubeSat applications. J Phys: Conf Ser (JPCS)
3. Lomaka I, Kramlikh A, Shafran S, Shklyar A (2021, Feb) Nanosatellite dipole antenna deployment mechanism. AIP Conf Proc 2318(1):180009)
4. "Tuna Can" antenna design for CubeSat missions. https://elfin.igpp.ucla.edu/s/TUNA-CAN-ANTENNA-DESIGN-FOR-CUBESAT-MISSIONS.pdf

Development of Payload Data Transmitter Using 8-bit Microcontroller and FM Transceiver for CubeSats

Rahul G. Waghmare, V. Suresh Kumar, K. R. Yogesh Prasad, Suman R. Valke, L. Suvarna, N. Ramalakshmi, and D. Venkataramana

Abstract This article presents a low cost and easy to realize payload/telemetry data transmitter for small satellite using commercial-off-the-shelf (COTS) components such as 8-bit microcontroller and FM transceiver module. 8-bit microcontroller handles the baseband modulation to generate audio signal which is then FM modulated on RF carrier frequency. The prototype model, developed using the design presented in this paper, supports data rate of 1200 baud rate using audio frequency shift keying (AFSK) at baseband and frequency modulation (FM) at radio frequency in VHF band. Since the baseband modulation is handled by 8-bit microcontroller, data rate and modulation schemes can be implemented by software, thereby making the design more flexible and reconfigurable. Moreover, since the RF front end is supported by FM transceiver, the same system can be extended to receive the tele-commands from ground in half duplex mode, provided that the baseband demodulation is handled by the microcontroller.

Keywords Data transmitter · Microcontroller · Baseband modulation · Satellite communication · APRS Digipeater · Slow scan television

1 Introduction

Building a small or nano satellite for a low earth orbit presents several challenges, especially for institutions with limited funding or resources. The conventional approach of satellite building includes use of Hi-Rel space grade components which are often outrageously expensive and requires significant amount of expertise. Keeping this in mind, we have developed a transmitter using commercial-off-the-shelf (COTS) components for housekeeping data as well as payload data transfer from satellite to ground. The transmitter is designed using 8-bit microcontroller for processing of baseband and FM transmitter for modulation of baseband signals over radio frequency.

R. G. Waghmare (✉) · V. Suresh Kumar · K. R. Yogesh Prasad · S. R. Valke · L. Suvarna · N. Ramalakshmi · D. Venkataramana
U R Rao Satellite Centre, ISRO, Bangalore 560 017, India
e-mail: rahulgw@ursc.gov.in

D. Singh et al. (eds.), *Computer Aided Constellation Management and Communication Satellites*, Lecture Notes in Electrical Engineering 987,
https://doi.org/10.1007/978-981-19-8555-3_4

Since the beginning, CubeSats have been proving their applications in many space-based scientific experiments for educational institutions. Till date, around one thousand CubeSats missions have been launched across the globe, and the numbers are increasing day by day. Most CubeSats employ VHF/UHF/S band frequencies for their RF link between satellite and ground station [1–3]. Various modulation schemes are being developed using digital techniques for efficient and reliable transmission such as GMSK [4, 5], Audio FSK [6], FSK [7], and many more [8–10]. These modulations can be easily developed for the 8-bit microcontroller using open-source development environments such as Arduino IDE.

In this paper, we present overall details of the data transmitter for satellite application with two case studies where we used the same design to adopt two different baseband modulation schemes, namely audio frequency shift keying (AFSK) for transmission of automatic packets reporting system (APRS) packets and frequency modulation (FM) for slow scan television (SSTV) images in PD120 format. Section 2 represents the overall design of the transmitter; Sect. 3 represents the link budget calculation for SDR-based ground station. Sections 4 and 5 provide the case studies of the transmitter used as APRS Packet Digipeater and SSTV transmission, respectively.

2 Design Methodology

The data transmitter is built around FM transceiver which acts as a RF front end to perform modulation and demodulation. Input and output of the FM transceiver are analog audio, which is then processed by an 8-bit microcontroller. Here, analog audio acts as an IF/baseband signal which implements digital modulation at baseband level using microcontroller. Since the microcontroller can be programmed for customized modulation schemes, this makes the proposed design truly versatile.

2.1 Structure of the Transmitter

Major blocks of the data transmitter are shown in Fig. 1. The data transmitter constitutes of four main modules, namely power supply module, RF transceiver, baseband modulator, and SPI flash memory.

Power Supply Module

The data transmitter receives power from unregulated raw bus of the satellite and generates 5 V supply for microcontroller (Atmega328P) and 4 V for transceiver (SA818v) using separate stepdown converters (LT1936). LT1936 provides cycle-by-cycle current limit, frequency fold-back, and thermal shutdown that provides protection against shorted outputs, and soft-start eliminates input current surge during start-up.

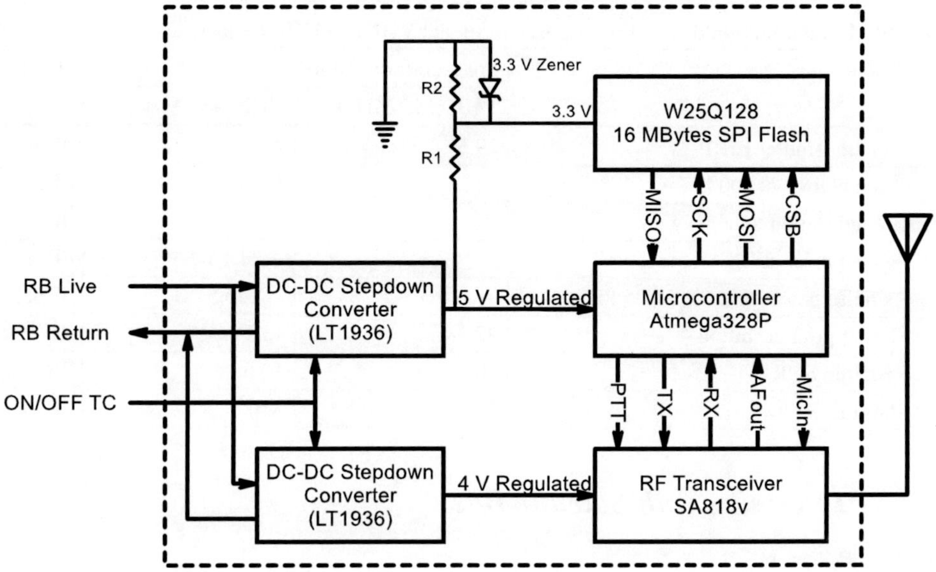

Fig. 1 Block diagram of the data transmitter

FM Transceiver

The FM transceiver, SA818v, operates on frequency range from 134 to 174 MHz. It can be configured for a single channel of either 12.5 kHz or 25 kHz with maximum frequency deviation of 2.4 kHz. The FM transceiver needs to be configured over UART interface, and the output power can be controlled using level command either in high (1 W) or low (0.5 W) power mode. The transceiver can be kept in sleep mode by using a level command. These functionalities are implemented in the proposed design by a microcontroller.

Baseband Modulator

Baseband modulator for converting digital data into audio is built around Atmega328P 8-bit microcontroller. This microcontroller runs on 16 MHz clock and 5 V power supply. It has 32 KB of flash-based program memory, 1 KB of EPROM, and 2 KB of SRAM. It supports 8 channel 10-bit ADC and 23 programmable I/O lines. These resources make this microcontroller ideal for the required digital signal processing.

Memory Module

SPI flash-based memory is interfaced with microcontroller as a secondary memory. This memory can be used to store the pre-recorded data for broadcasting purpose or the data received from payloads for transmissions upon ground station visibility. Interface between OBC and data transmitter can be used to transfer the data from payload to SPi flash during the non-availability of the ground station.

Table 1 Link margin details for proposed design at VHF and UHF frequencies

Parameters	Spacecraft downlink		Units
	VHF (145 MHz)	UHF (435 MHz)	
Data transmitter EIRP	29.5	29.5	dBm
Channel loss @ 500 km Alt, 10° Elev	146.3	155.8	dB
Ground station antenna gain	14	14	dBi
Signal power at LNA of receiver	– 102.8	– 112.3	dBm
Rx noise power @ 650 °K, 25 kHz	– 126.5	– 126.5	dBm
SNR at ground station	22.7	13.1	dB
Required SNR	10.6	10.6	dB
Margin	12.1	2.5	dB

2.2 *Interfaces with Satellite Bus*

Data transmitter is designed to be a single card solution on PCB of standard (PC-104) dimensions so that it can be used across various nanosatellite buses. Card has an SMA connector for RF and a 9-pin connector for power, telemetry and tele-command interface.

The system is designed to meet the low-power consumption requirements of the small/nanosatellites. Total power consumption of the proposed design is around 3.2 W during transmit mode and 0.75 W during non-transmit mode.

3 Link Budget Calculations

See Table 1.

4 Case Study I: APRS Digipeater

APRS is short for automatic position/packet reporting system, which was designed by Bob Bruninga, and introduced at the 1992 TAPR/ARRL Digital Communications Conference. Fundamentally, APRS is a packet communications protocol for disseminating live data to everyone on a wireless network in real time. It uses AX.25 unnumbered information (UI) frames at data link layer and AFSK/FM with NRZi encoding at physical layer. Each packet is appended with frame checksum to ensure the error detection procedure. For further details, see [11].

Longer distance packet communication in APRS network is achieved by APRS Digipeater. APRS Digipeater is the digital repeater which receives the APRS packets and retransmit them on the same frequency. Since FM transceiver is used for RF front end, the proposed design can be easily adopted to receive the ground commands, provided that the decoding of the commands is handled by microcontroller software. For APRS Digipeater case study, the software for processing of APRS signals is written using Arduino IDE. This software performs the following tasks:

- Configures FM transceiver for receiving as well as transmitting on 145.825 MHz
- Samples the incoming audio at 9600 samples per second
- Demodulates AFSK signal using Goertzel's algorithm
- Decodes of AX.25 frame
- Generates and verifies the CRC checksum of the APRS packet
- Searches for its address or alias among Digipeater addresses, sets the corresponding bit for 'digipeated' in the respective byte
- Forms AX.25 frame using the decoded information and modulate it over AFSK
- Transmits the frame by enabling the push-to-talk (PTT) of the transceiver.

The test setup was rigged, as shown in Fig. 2, to validate the functionalities of APRS Digipeater. Test setup consists of an open platform, UISS, for processing the APRS packets, signal generator for FM modulation, and SDR-based receiver to receive the digipeated packets. During the testing of Digipeater performance, UISS was kept to generate APRS packets at an interval of 10 s. These generated packets (analog audio) are fed to signal generator for FM modulation over carrier frequency. Since the proposed design has a common interface for RF input and output, external RF switch is required to de-multiplex them. The output of the RF switch (i.e., output of the Digipeater) is then attenuated and fed to SDR which performs I-Q sampling on the RF. These I-Q samples are then processed on computer using SDR# software to FM demodulate and generate the audio signal. These Audio FSK signals are then processed by UISS and compared with the transmitted packets. Digipeater efficiency is the ratio of no. of packets transmitted by UISS to the no. of digipeated packets received by UISS. Digipeater efficiency of 100% was observed up to the received signal strength of − 110 dBm in response to 100 packets transmitted by UISS.

5 Case Study II: SSTV Image Transmitter

Slow scan television (SSTV) image transmission is an analog mode of picture transmission where images are scanned line by line and converted into FM modulated audio signal of frequencies between 1100 and 2300 Hz. Each pixel value is represented by a unique frequency between 1500 Hz (for 0 level) and 2300 Hz (for 255 level). There are more than dozen protocols for converting image into audio and retrieving back the image. Readers are directed to [12] for further details.

For the case study presented in this article, we have chosen PD120 SSTV protocol that optimizes the transmission time and preserves colors of the image. Color images of 640 × 480 pixels were stored onto SPI flash memory, and the software was developed in Arduino IDE to read the image and to implement the SSTV algorithm. Figure 3 depicts the details of generic SSTV transmission protocols.

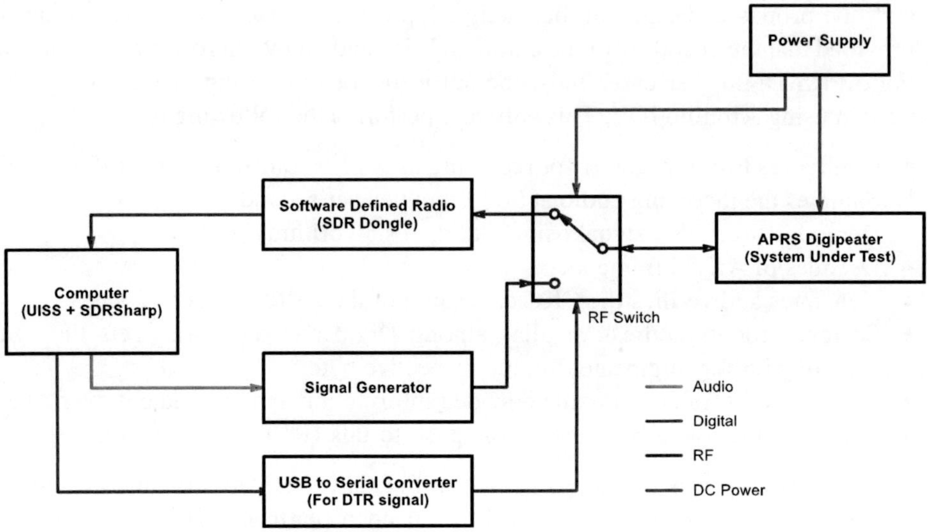

Fig. 2 Block diagram of test setup

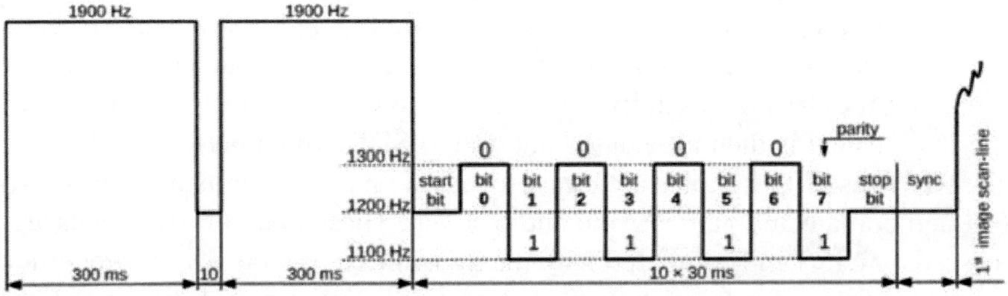

(a) Format of the header/synchronization pattern for SSTV signals

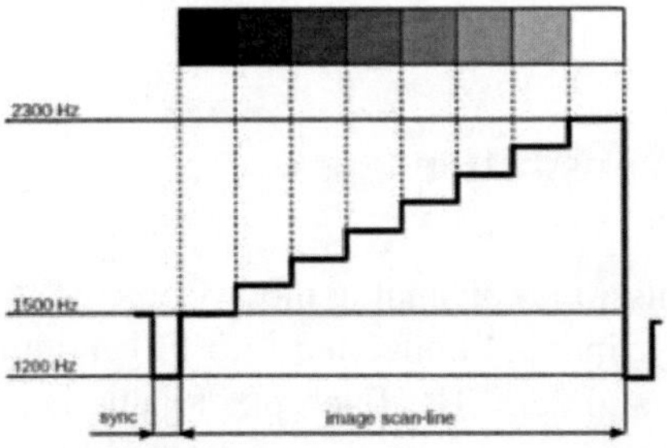

(b) Mapping function [Pixel value → freq] (c) Received SSTV image

Fig. 3 Details of the SSTV image transmission protocols

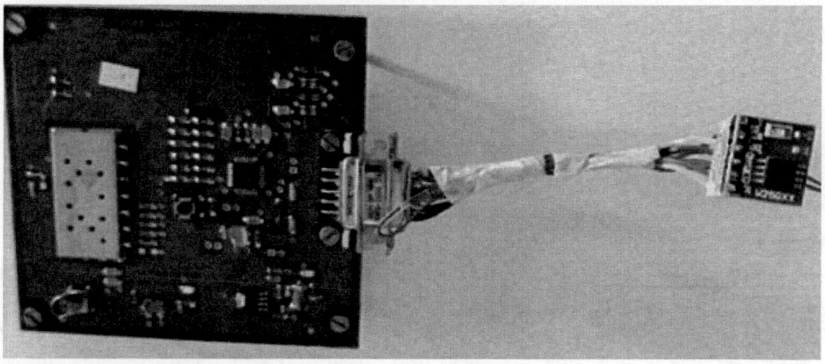

Fig. 4 SSTV image transmitter

SDR-based receiver was used to receive the FM signal from data transmitter, and an open platform (RXSSTV) was used to decode the SSTV image. Figure 3c shows the received image after decoding by RXSSTV software. Figure 4 shows the proposed system configured for SSTV image transmission.

6 Conclusions and Future Scope

In this article, we have presented a generic and reconfigurable data transmitter design to be used for small and nanosatellites. This design can be easily modified for robust communication between satellite and ground station by developing customized modulation and encoding algorithms using 8-bit microcontroller programs. Also, transmitter design uses FM transceiver, thereby making it useful for receive applications in half duplex mode. Since the design includes SPI flash memory, the following is the list of applications one can implement using the proposed design:

- Store the payload data such as images, time series data, etc., and transmit upon ground station visibility
- Store pre-recorded images, audio files before launch and broadcast them to public to commemorate certain historical event
- Use as a telemetry and tele-command system in half duplex mode with periodical TM bursts followed by receive mode.
- Although the FM transceiver used in proposed design works in VHF, pin compatible FM transceiver (SA818u) can also be used without changing the PCB design. This will enable system to work in UHF band.

References

1. Gomx-3. https://gomspace.com/gomx-3.aspx. Accessed: 25 March 2022
2. Dice. https://earth.esa.int/web/eoportal/satellite-missions/content/-/article/dice
3. Quakesat. https://directory.eoportal.org/web/eoportal/satellite-missions/q/quakesat
4. Lee J et al (2011) Implementation of a GMSK communication system on FPGA. In: 2011 IEEE Second Latin American symposium on circuits and systems (LASCAS). IEEE
5. Svedek T et al (2009) A simple signal shaper for GMSK/GFSK and MSK modulator based on sigma-delta look-up table. Radio-Engineering 18(2009):230–237
6. Daffalla MM (2013) Design and analysis of low baud-rate modem using microcontrollers. In: 2013 International conference on computing, electrical and electronic engineering (ICCEEE), pp 44–48
7. Menna BV, Acosta GG, de la Vega RJ (2015) Low cost programmable FSK modem. In: 2015 XVI workshop on information processing and control (RPIC), pp 1–6
8. Hassan DMA, Corral-De-Witt D, Ahmed S, Tepe K (2018) Narrowband data transmission in TV white space: an experimental performance analysis. In: 2018 IEEE international symposium on signal processing and information technology (ISSPIT), pp 192–196
9. Harris JS (2016) Analysis and implementation of communications systems for small satellite missions. Master of Science (MS), Thesis, Electrical & Computer Engineering, Old Dominion University
10. Zeedan A, Khattab T (2022) A critical review of baseband architectures for CubeSats communication systems. arXiv, 2022. https://arxiv.org/abs/2201.09748
11. The APRS Working Group. APRS protocol reference protocol version 1.0. Available at: http://www.aprs.org/doc/APRS101.PDF
12. Bruchanov M. Image communication on short waves. Available at: https://www.sstv-handbook.com/download/sstv-handbook.pdf

Dual-Band Terahertz Metamaterial Absorber for a Sensor Application

Laxmi Narayana Deekonda, Sanjay Kumar Sahu, and Asit Kumar Panda

Abstract This paper includes a metamaterial design at THz range, explicitly useful as an absorber. It is composed of three layers with top and bottom are to be gold and separated by a dielectric material in between. The structure provides 100% at a lower frequency of 0.846 THz and 98.6% at a higher frequency of 2.12 THz. The proposed Complementary Split Ring Resonators (CSRR) shows a tuning range from 2.12 to 2.16 THz. The CSRR parameter is optimized to get dual resonance frequency. The proposed structure provides a quality factor of 19.2 at a lower band frequency and 19.56 at an upper band frequency.

Keywords Dual-band · Absorber · Terahertz · CSRR · Inter-satellite links

1 Introduction

The significant advancement in the field of metamaterial brings the research into the mainstream of sciences, especially electromagnetics [1]. There are numerous applications found such as cloaking, scattering cancelation, filtering, imaging and many others. Recently, it is extended to space industry pertaining to satellite application as many private and government agencies involve in space missions [2]. Out of those absorbers is one of the areas where this technique is widely used. The metamaterial absorbers at a sub-wavelength scale are used in defense and civil applications. The first metamaterial absorber design used the metallic electric ring resonator in 2008 [3]. In this design, FR4 substrate is used as a substrate that is ultra-thin compared to a convention absorber, and due to this important feature, metamaterial absorbers have been used in elective energy measurement, spectrum imaging, thermal/refractive

L. N. Deekonda (✉) · S. K. Sahu
Department of ECE, Lovely Professional University, Phagwara, Punjab, India
e-mail: laxminarayana.lpu@gmail.com

S. K. Sahu
e-mail: sanjay.23393@lpu.co.in

A. K. Panda
Department of ECE, National Institute of Science and Technology, Berhampur, Orissa, India

D. Singh et al. (eds.), *Computer Aided Constellation Management and Communication Satellites*, Lecture Notes in Electrical Engineering 987,
https://doi.org/10.1007/978-981-19-8555-3_5

index sensing, detection, etc. Recently, metamaterial absorbers are used as ultra-thin sensors/biosensor with narrow bands [4]. In the metamaterial-based sensor design, main challenges are high Q-factor, polarization insensitive and independent incident angle [5]. In [5], two metal rings with the cylinder have proposed to provide dual-band absorption and high Q-factor but the allowed incident angle is only 30°. The two rectangular rings and a circle are used in a design with silver to obtain a dual-band perfect absorber which provides a high Q-factor [6]. But in this structure, three resonators are used for the dual-band. In [7], simple cross slot is designed to realize an absorber for biosensing. Now people are moving toward optical sensing for biomedical research [8]. In [8], microfluidic-based sensor has been designed but it gives a low-value Q-factor.

In this paper, rectangular Complementary Split Ring Resonators (CSRR)-based metamaterial has been designed to get dual-band operation. The gold metal used to design CSRR provides good absorption. This monolayer structure has two bands frequency 0.845 and 2.12 THz. This structure is designed using CST Microwave studio.

2 Absorber Design

A unit cell of the structure is shown in Fig. 1. The structure has three-layer. In this structure, a substrate is cladded with metal in the top and bottom layers. A top layer is metal CSRR, which provides dual-band absorbance. The substrate used is polyimide with a dielectric constant, $\epsilon_r = 3.5$, and height is h. The metal used on top and bottom is gold with a high conductivity (4.09×10^7 S/m) and thickness of t and t_g, respectively.

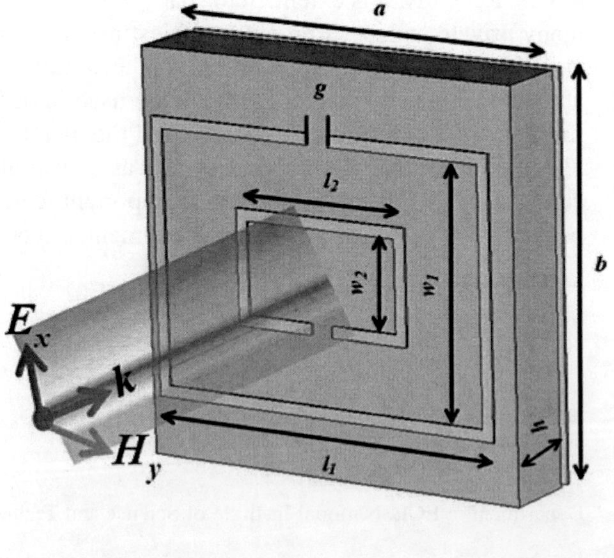

Fig. 1 Proposed unit cell based on CSRR

Table 1 Parameters

Symbols	Values (μm)	Symbols	Values (μm)
a	70	b	70
h	7	l_1	64
T	0.4	t_g	0.4
w_1	48	l_2	32
w_2	20	g	5

The unit cell size is $a \times b$ with a length of CSRR is l_1 and l_2. All the parameter values are mentioned in Table 1. The absorption, $A(\omega)$, of the metamaterial can be expressed by [1]:

$$A(\omega) = 1 - \left|R^2(\omega)\right| - \left|T^2(\omega)\right| \tag{1}$$

where $R(\omega)$ and $T(\omega)$ are S_{11} (i.e., the reflection coefficient) and S_{21} (i.e., the transmission coefficient), respectively. For the calculation $A(\omega)$, S_{11} and S_{21} are needed. The S-parameters of the structure are extracted from the electromagnetic simulator CST Microwave Studio. Due to the presence of metal (i.e., gold) in the bottom layer, so $T(\omega) = 0$. Because of this, above Eq. (1) is written as:

$$A(\omega) = 1 - \left|R^2(\omega)\right| \tag{2}$$

3 Results and Discussion

The proposed structure contains CSRR as a resonator. The simulated reflection spectrum is shown in Fig. 2. The figure showing proposed CSRR is resonating at two frequency 0.846 and 2.12 THz. Figure 3 shows corresponding absorption spectrum. The structure is perfect absorber at 0.846 THz frequency and 98.6% at frequency 2.12 THz.

The absorber performer analysis with variation of refractive index shown in Fig. 4. In this the value of refractive index, n is changing from 1 to 1.1 with the variation of 0.01. Figure 5 shows frequencies shift with respect to refractive index. It showing upper band has more frequencies shift compare to lower band frequency. The Q-factor of metamaterials is calculated formula [4]

$$Q = \frac{f_o}{\text{FWHM}} \tag{3}$$

Here, f_o is peak frequency and FWHM is full width at half maxima. The calculated Q-factor with change in refractive index is plotted in Fig. 6. The highest value Q-factor is 19.2 at a lower band frequency with a refractive index of $n = 1.1$ and the

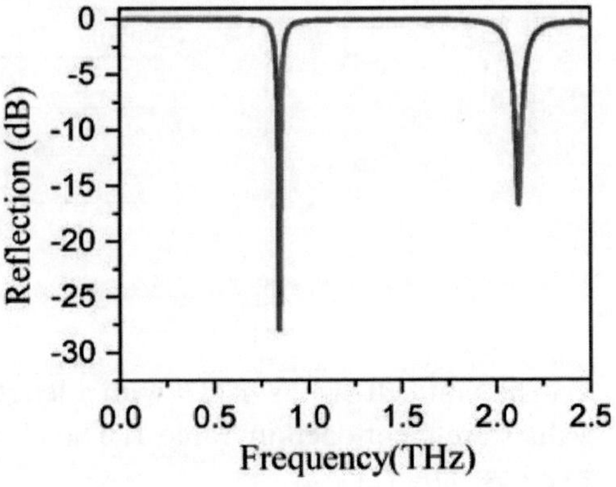

Fig. 2 Reflection spectrum of CSRR

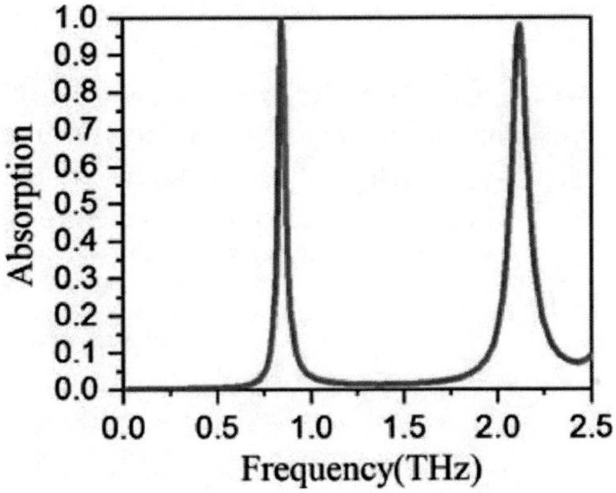

Fig. 3 Absorption spectrum of proposed CSRR

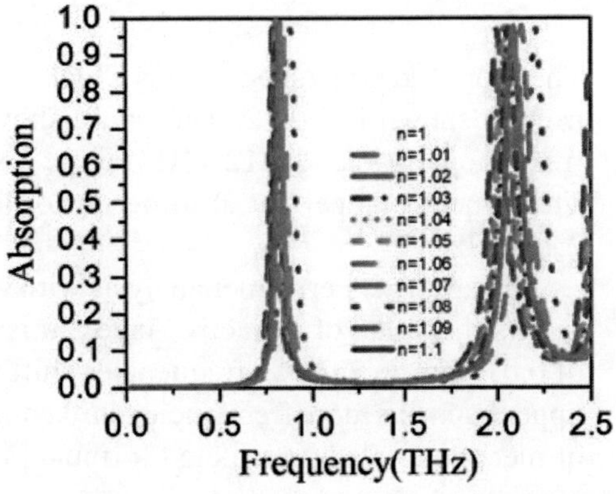

Fig. 4 Absorber performance analysis with variation of refractive index

upper band has 19.56 at a refractive index of $n = 1$. The sensitivity, S, of the proposed structure can be calculated by [6]

$$S = \frac{\Delta f}{\Delta n} \tag{4}$$

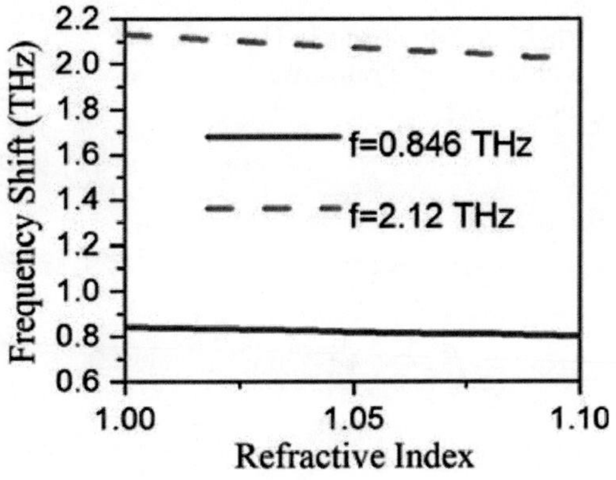

Fig. 5 Frequency shift analysis with variation of refractive index

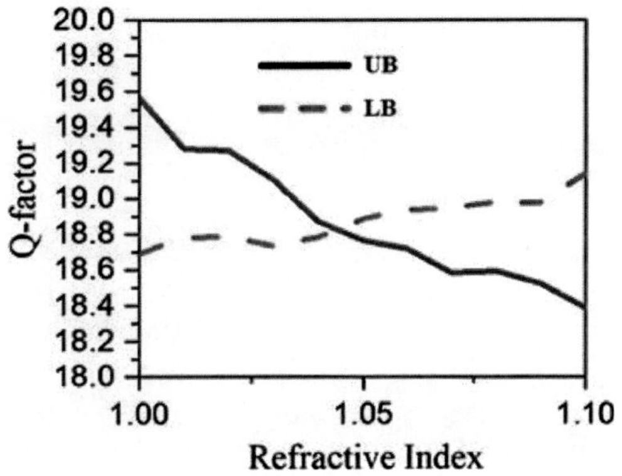

Fig. 6 *Q*-factor analysis with variation of refractive index

where Δn is change in refractive index and Δf is the frequency shift. Figure 7 is showing the sensitivity plot, which shows that lower band is less sensitive compared to upper band. In the upper band, value changes from to 1.3 THz/RIU with respect to refractive, and in case of lower band, 0.2 THz/RIU to 0.6 THz/RIU. The figure of merit (FOM) is calculate using sensitivity divided by FWHM. The figure of merits (FOMs) of proposed structure is 11.43 RIU-1 and 11.79 RIU-1 for LB and UB, respectively.

We made a comparison with existing literature to show the kind of improvement we obtained which is given in Table 2. The proposed design is dual band with high-quality factor and FOM is also high.

4 Conclusion

A CSRR-based metamaterial is designed which offers dual-band absorber. The absorption of structure is 100% and 98.6% in lower and upper band, respectively. But upper band frequency is more sensitive compare to lower band frequencies. The

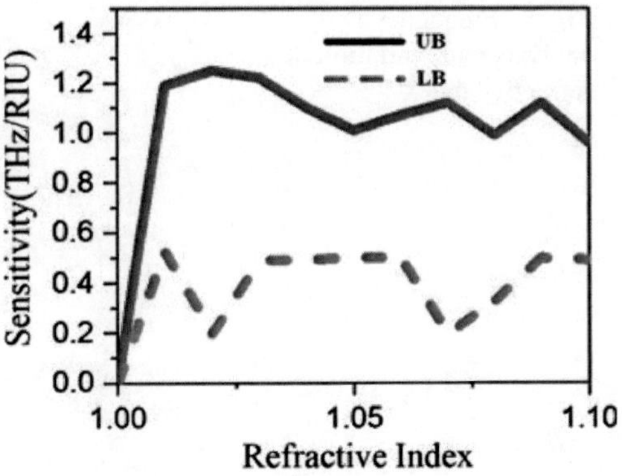

Fig. 7 Sensitivity analysis with variation of refractive index

Table 2 Comparison between available literatures

Reference	Q (THz/RIU)	FOM (RIU^{-1})
[7]	11.6	2.3
[8]	8.56	4
[9]	8.5	0.85
[10]	7	0.5
[11]	7.8	0.7
This work	19.2 and 19.56	11.43 and 11.79

proposed structure has quality factor of 19.20 THz/RIU and 19.56 THz/RIU. The FOM of structure is 11.43 RIU^{-1} and 11.79 RIU^{-1}.

References

1. Garcia JG, Martin F, Falcone F, Bonache J, Gil I, Lopetegi T, Laso MAG, Sorolla M, Marques R (2004) Spurious passband suppression in microstrip coupled line band pass filters by means of split ring resonators. IEEE Microw Wirel Comp Lett 14(24):416–418
2. Civas M, Akan OB (2021) Terahertz wireless communication system. ITU J Future Evolving Technol 2(7):5
3. Wang BX, Zhai X, Wang GZ, Huang WQ, Wang LL (2015) Design of a four-band and polarization-insensitive terahertz metamaterial absorber. IEEE Photonics J 7(1):1–8
4. Landy NI, Sajuyigbe S, Mock JJ, Smith DR, Padilla WJ (2008) Perfect metamaterial absorber. Phys Rev Lett 100:207402
5. Yan X, Liang L-J, Ding X, Yao J-Q (2017) Solid analyte and aqueous solutions sensing based on a flexible terahertz dual-band metamaterial absorber. Opt Eng 56(2):027104
6. Janneh M, De Marcellis A, Palange E, Tenggara AT, Byun D (2018) Design of a metasurface-based dual-band Terahertz perfect absorber with very high Q-factors for sensing applications. Optics Commun 416:152–159
7. Cong L, Tan S, Yahiaoui R, Yan F, Zhang W, Singh R (2015) Experimental demonstration of ultrasensitive sensing with terahertz metamaterial absorbers: a comparison with the metasurfaces. Appl Phys Lett 106:031107

8. Hu X, Xu G, Wen L, Wang H, Zhao Y, Zhang Y, Cumming DRS, Chen Q (2016) Metamaterial absorber integrated microfluidic terahertz sensors. Laser Photonics Rev 10:962
9. Li Y, Chen X, Hu F, Li D, Teng H, Rong Q, Zhang W, Han J, Liang H (2019) Four resonators based high sensitive terahertz metamaterial biosensor used for measuring concentration of protein. J Phys D: Appl Phys 52:095105
10. Islam M, Rao SJM, Kumar G, Pal BP, Chowdhury DR (2017) Role of resonance modes on terahertz metamaterials based thin film sensors. Sci Rep 7:7355
11. Saadeldin AS, Hameed MFO, Elkaramany EMA, Obayya SSA (2019) Highly sensitive terahertz metamaterial sensor. IEEE Sens J 19:7993

Studying the Applications of Graph Labeling in Satellite Communication Through 2-Odd Labeling of Graphs

Ajaz Ahmad Pir, Tabasum Mushtaq, and A. Parthiban

Abstract The necessity for good and tenable communication systems has motivated researchers to develop mobile communication networks (MCN). On the other hand, the huge functionalities of the global system of mobile (GSM) communication have given an increasing number of users. As the subscribers grow, the necessity for efficient and productive planning of the limited frequency spectrum of the GSM is inevitable, especially in highly dense areas. Researchers have proposed various algorithms for frequency or channel allocation (CA), as the discussions about CA methods to resolve the various practical issues in CA are going on. The literature reveals that the "Manhattan distance" concept can be used in scheduling and optimization problems. Similarly, the same idea makes it possible to discover a more tenable telecommunication system with "ease of connectivity" among subscribers, even when many users are on a common channel. Graph labeling is the most interesting idea in graph theory that has numerous uses in different fields, particularly in communication networks. 2-odd labeling assigns distinct integers to the nodes of a graph $G(V, E)$ such that the positive difference of adjacent nodes is either 2 or an odd integer, $2k \pm 1,\ k \in N$. The motivation behind the development of this article is to study the applications of graph theory in communication networks through the concept of 2-odd labeling in graphs.

Keywords Graph labeling · Communication network · 2-odd labeling · Degree splitting graph of a given graph

A. A. Pir
Department of Mathematics, Lovely Professional University, Phagwara, Punjab 144 411, India

T. Mushtaq
Department of Statistics, Government Degree College Sopore, Sopore, Jammu and Kashmir 193 201, India

A. Parthiban (✉)
Mathematics Division, School of Advanced Sciences, Vellore Institute of Technology, Vellore, India
e-mail: parthiban.a@vit.ac.in

D. Singh et al. (eds.), *Computer Aided Constellation Management and Communication Satellites*, Lecture Notes in Electrical Engineering 987,
https://doi.org/10.1007/978-981-19-8555-3_6

1 Introduction

A graph labeling is a function that assigns integers to the lines or nodes, or both of G under some conditions. The importance of graph labeling includes its numerous applications in many areas like circuit design, radar, communication network address, etc. Network representations are playing a vital role in various aspects of computer science and communication networks. In order to recover important data, one has to access a global information structure representing the whole network. The "diagrammatic representation" of the graph model applied on these data sets gives strong support to visualizing and understanding the data. For a detailed study, see [1–5].

The graphs considered are simple, finite, undirected, and connected. Let N, P, and Z, respectively, be the set of all natural numbers, primes, and integers. 2-odd labeling of $G(V(G), E(G))$ is a 1–1 function $g : V(G) \rightarrow Z$ such that the positive difference between every pair of adjacent nodes x_1 and x_2, i.e., $|g(x_1) - g(x_2)|$ is either $2k \pm 1$; $k \in N$ or exactly 2. For more results on 2-odd graphs, see [6–9].

1.1 Preliminaries

A few relevant definitions and results are recalled in this section.

Definition 1 A path of length r, P_r, from u_0 to u_r in G is a sequence of $r \geq 0$ lines $e_i : 1 \leq i \leq r$ of G such that $e_i = \{u_{i-1}, u_i\}$ [10].

Definition 2 If in G every distinct pair of nodes are adjacent, then it is a complete graph K_n [5].

Definition 3 The wheel $W_s = C_{s-1} \wedge K_1$ on s nodes $s \geq 4$, constructed by connecting K_1 to every node of C_{s-1} [11].

Definition 4 A helm graph H_s is formed by inserting a line and node to each rim node of W_s [12].

Definition 5 A fan $F_{1,r}$ is defined as $K_1 \wedge P_r$ [13].

Definition 6 The Dutch windmill D_r or friendship graph has r copies of K_3 joined at the common node w_0 [7].

Definition 7 The generalized butterfly graph, BF_n, is formed by attaching nodes to each wing with the property that the sum of attaching nodes to each wing is the same [14].

Definition 8 Consider H with $V = S_1 \cup S_2 \cup \ldots \cup S_i \cup T$, where each S_i is a set of nodes with minimum two nodes of the same degree and $T = V \cup S_i$ [15]. The degree splitting graph of H, DSG(H) is derived from H by inserting nodes $w_i : 1 \leq i \leq t$ and connecting to every node of S_i for $1 \leq i \leq t$.

Theorem 1 *Every subgraph of a 2-odd graph is 2-odd* [7].

Theorem 2 W_{n+1} *is 2-odd* [6].

Proposition 1 K_n, $n \geq 5$ *is not 2-odd* [7].

In this paper, 2-odd labeling of degree splitting graphs of $P_r, K_n, W_s, H_s, F_{1,r}, D_r, \mathrm{BF}_n$, *are obtained. Let WLG denote without loss of generality.*

2 Results and Discussion

This section is dedicated to deriving 2-odd labeling of DSG(P_r), DSG(K_n), DSG(W_s), DSG(H_s), DSG($F_{1,r}$), DSG(D_r), and DSG(BF_n).

Theorem 3 DSG(P_n) *admits 2-odd labeling.*

Proof Take P_n; $n \geq 2$, and obtain DSG(P_n) with $V(\mathrm{DSG}(P_n)) = V_1 \cup V_2$, where $V_1 = V(P_n)$ and $V_2 = \{x, y\}$ and the nodes of degree 1 and 2 are joined with x and y, respectively, (see Fig. 1). Clearly, , $|DSG(P_n)| = n+2$. Define $g : V(DSG(P_n)) \rightarrow Z$ as follows: WLG, let $g(x) = 2k$; $k \in N$, $g(y) = 0$, and $g(u_i) = 2i-1 : 1 \leq i \leq n$. Hence, g is 2-odd labeling of DSG(P_n).

Lemma 1 DSG(C_n) $= W_{n+1}$, *for* $n \geq 3$.

Proof Take C_n with $v_1, v_2, \ldots, v_n$ and obtain DSG(C_n). Since the degree of every node in C_n is 2, the newly introduced node, say w, is adjacent to all the nodes of C_n. Thus, w becomes the hub of the resultant graph W_{n+1} (see Fig. 2).

Theorem 4 *If a graph H is 2-regular, then* DSG(H) *admits 2-odd labeling.*

Proof The proof is direct from Lemma 1 and Theorem 2.

Lemma 2 DSG(K_n) $= K_{n+1}$.

Proof Take K_n and obtain DSG(K_n). Since every node is of the same degree, the newly introduced node is joined to all the nodes of K_n and hence forms K_{n+1}.

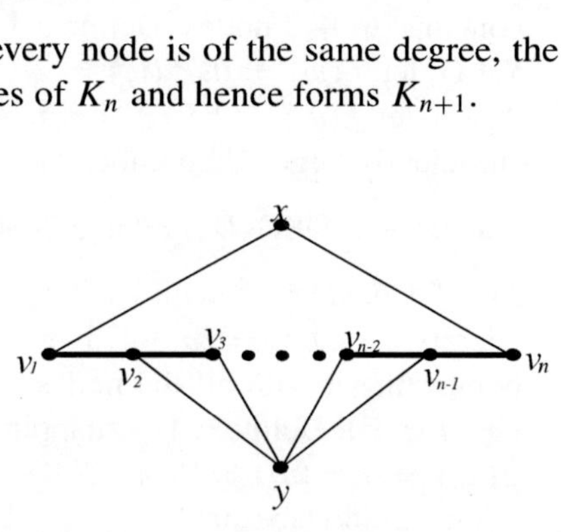

Fig. 1 DSG(P_n)

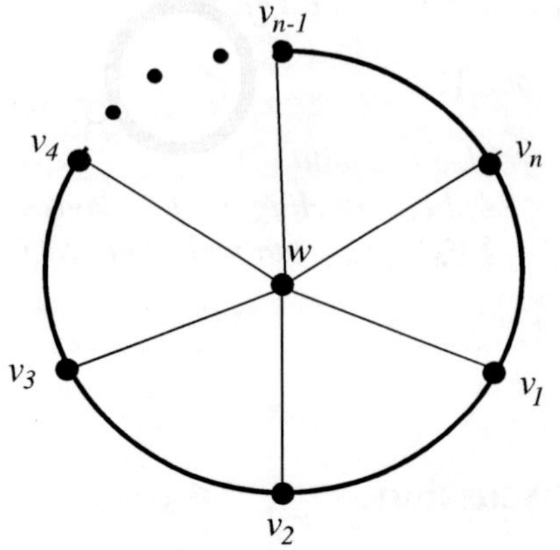

Fig. 2 DSG(C_n)

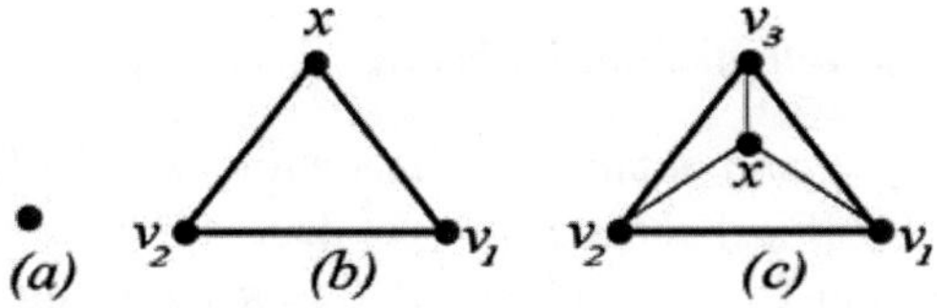

Fig. 3 DSG(K_i) : $1 \leq i \leq 3$

Remark 1 One can see DSG(K_i) : $1 \leq i \leq 3$ in Fig. 3 and obtain 2-odd labeling.

Theorem 5 DSG(K_n) *does not admit 2-odd labeling* $\forall n \geq 4$.

Proof The proof is direct from Lemma 2 and Proposition 1.

Theorem 6 DSG(H_n) *admits 2-odd labeling* $\forall n \geq 4$.

Proof Let H_n be the helm on $2n + 1$ nodes, namely $V(H_n) = \{v_1, v_2, \ldots, v_n, u_1, u_2, \ldots, u_n, w\}$, where w is the central node. Obtain DSG(H_n) by introducing two new nodes, say x and y. Join x with the nodes of degree 4 and y with the nodes of degree 1 (see Fig. 4). One can see that the resultant graph DSG(H_n) contains $2n + 3$ nodes. Define a 1–1 mapping $f : V(\text{DSG}(H_n)) \rightarrow Z$ as follows: WLG, let $f(w) = 0$, $f(v_i) = 2i - 1; 1 \leq i \leq n - 1$, $f(v_n) = 2$, and $f(x) = 4$. Again, let $f(u_i) = 2i; 1 \leq i \leq n - 1$, $f(u_n) = t$, and $f(y) = t + 2$, where t is any sufficiently large odd number. Hence, DSG(H_n) admits 2-odd labeling for $n \geq 4$.

Theorem 7 DSG(D_n) *admits 2-odd labeling.*

Proof Consider D_n with $V(D_n) = \{u_0, u_1, \ldots, u_{2n}\}$, where u_0 is the common node. Clearly, $|V(D_n)| = 2n + 1$. Now obtain the DSG(D_n) by introducing a node x and connecting it with all the nodes of degree 2 and so $|V(\text{DSG}(D_n))| = |V(D_n)| + 1$ (see Fig. 5). Define a 1–1 mapping $g : V(\text{DSG}(D_n)) \rightarrow Z$ as follows: WLG, let $g(u_0) = 0$, $g(u_i) = 2i + 1$; $1 \leq i \leq 2n$, and $g(x) = 2t; t \in N$. Thus, DSG(D_n) admits 2-odd labeling.

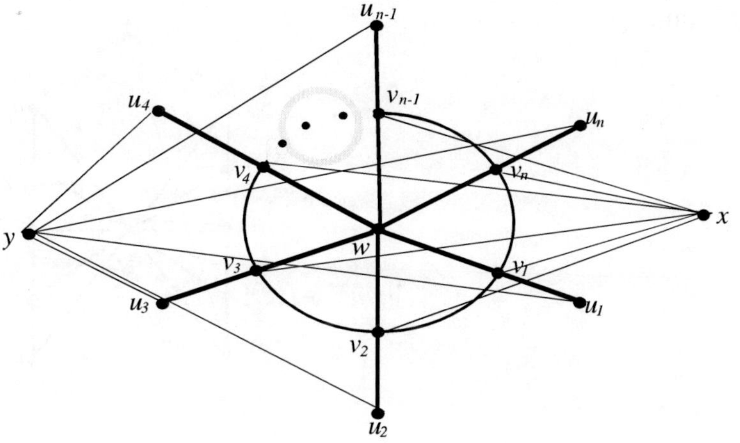

Fig. 4 DSG(H_n)

Fig. 5 DSG(D_n)

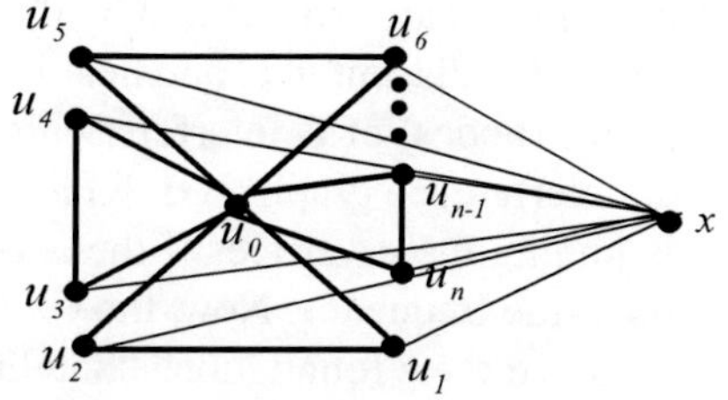

Theorem 8 DSG(BF_n) *admits 2-odd labeling* $\forall n \geq 2$.

Proof Take BF_n on $2n + 1$ nodes, namely $V(\mathrm{BF}_n) = \{u_0, u_1, ..., u_n, v_1, v_2, ..., v_n\}$, where u_0 is the common node. So BF_n has $2n$ copies of C_3 joined at u_0. Clearly, $|V(\mathrm{BF}_n)| = 2n+1$. Obtain DSG($\mathrm{BF}_n$) by introducing two nodes x and y and joining x with all the nodes of degree 2 and y with the nodes of degree 3 (see Fig. 6). Clearly, $|V(\mathrm{DSG}(\mathrm{BF}_n))|=|V(\mathrm{BF}_n)| + 2$. Define a 1–1 mapping $g : V(\mathrm{DSG}(\mathrm{BF}_n)) \rightarrow Z$ as given: WLG, let $g(u_0) = 1,\ g(u_i) = 2i; 1 \leq i \leq n,\ g(v_1) = g(u_n) + 2,\ g(v_i) = g(v_{i-1}) + 2; 2 \leq i \leq n, g(x) = 2t - 1; t \in N$, and $g(y) = r$, where r is any sufficiently large unused odd number. Thus, DSG(BF_n) admits 2-odd labeling.

3 Applications of Graph Labeling in Communication Networks

To find a productive way, "safe transmissions" are required in places like wireless telephony, Wi-Fi, security systems, etc. Two close channels can interfere, thereby harming communications, and this may be solved by means of a "suitable channel

Fig. 6 DSG(BF_n)

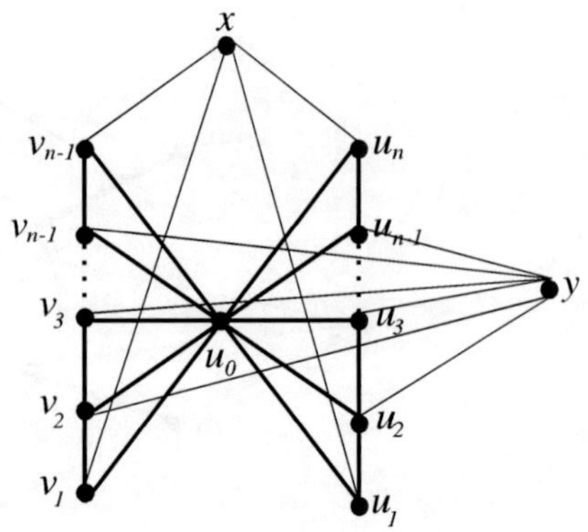

assignment." The "channel assignment problem" (CAP) is a task of assigning a frequency—non-negative integer, to each television or radio transmitter placed in different areas such that "communications" do not interfere. Graph theoretically, the transmitters are represented by the nodes of G; 2 nodes, say x and y are very near if x and y are adjacent in G and near if x and y are at distance of 2 apart in G. If the access points (nodes) of G interfere with other nodes in the same face, then G is known as an interference graph (IG). IG is obtained by the access points as vertices and a line is joining these vertices if the access points interfere with one another while using the same frequency. Now, the CAP is changed into a "graph labeling problem" [4].

Since their foundation, "Satellite Communications" (SatComs) have had many applications such as media broadcasting, backhauling, and newsgathering. Recently, following the development of "Internet-based applications," SatComs is undergoing a transformation stage by recasting the system structure on data services. The core reason is (1) the swift adoption of media streaming instead of "linear media broadcasting" and (2) the emergency requirement to extend "broadband coverage" to underserved places. For a detailed information, refer to [16].

Graph coloring is used in many research fields of computer science. For instance, a data structure can be modeled in the form of a tree which in turn used nodes and lines. Similarly, graph coloring is also used in resource allocation and scheduling. Moreover, some models in graph theory such as paths, walks, and circuits are also used in traveling salesman problems, database design concepts, and resource networking [17, 18].

4 Conclusion

In this present study, a few results concerning 2-odd labeling in the context of degree splitting graph of the given graph are explored. This study may lead to the complete characterization of 2-odd labeling which is still an open problem. The applications of graph labeling in communication networks and satellite communication are also highlighted. We also believe that the concept of 2-odd labeling may find its applications in space-related research.

References

1. Acharya BD, Arumugam S, Rosa A (2008) Labeling of discrete structures and applications. Narosa Publishing House, New Delhi
2. Bloom GS, Glomb SW (1977) Application of numbered undirected graphs. Proc IEEE 65:562–570
3. Gross J, Yellen J (1999) Graph theory and its applications. CRC Press, London
4. Prasanna NL (2014) Applications of graph labeling in communication networks. Int Res J Comp Sci Technol 7(1)
5. West DB (2000) Introduction to graph theory, 2nd edn. Prentice-Hall, Englewood Cliffs, NJ
6. Abirami P, Parthiban A, Srinivasan N (2020) On 2-odd labeling of graphs. Eur J Mole Clin Med 07:3914–3918
7. Laison JD, Starr C, Walker A (2013) Finite prime distance graphs and 2-odd graphs. Discret Math 313:2281–2291
8. Parthiban A, Gnanamalar David N (2017) On prime distance labeling of graphs. In: Arumugam S, Bagga J, Beineke L, Panda B (eds) Theoretical computer science and discrete mathematics. ICTCSDM 2016. Lecture notes in computer science 10398. Springer, Cham, vol 31, pp 238–241
9. Parthiban A, Pir AA, Felix A (2020) Certain results on prime and prime distance labeling of graphs. J Phys Conf Ser 1531:012062, 1–6
10. Gallian JA (2014) Graph labeling. Electron J Combin (Dyn Surv DS6) 17
11. Parthiban A, Samdanielthompson G, Kumar KS (2021) On finite prime distance graphs. Indian J Pure Appl Math 52:22–26
12. Meena S, Vaithilingam K (2013) Prime labeling for some helm-related graphs. Int J Inno Res Sci Eng Technol 2(4):1075–1085
13. Vaithilingam K (2014) Difference labeling of some graph families. Int J Math Statistics Invention (IJMSI) 2(6):37–43
14. Wahyuna HD, Indriati D (2018) On the total edge irregularity strength of generalized butterfly graph. J Phys Conf Ser 1008:1–6
15. Sampathkumar E (1980) Walikar: on the splitting graph of a graph. The Karnataka Univ J Sci 25:13–16
16. Kodheli O et al (2020) Satellite communications in the new space era: a survey and future challenges. IEEE Commun Surv Tutorials (DRAFT) 2:1–45
17. Tosuni B (2015) Graph coloring problems in modern computer science. Eur J Interdiscip Stud 01(2):87–95
18. Leighton FT (1979) A graph coloring algorithm for large scheduling problems. J Res Natl Bureau Standard 84:79–100

Application Scenario of Blockchain Security in Massive MIMO

Abdullah Mohammed and Shakti Raj Chopra

Abstract Blockchain, which is critical to the Satellite and other networks, has recently acquired popularity. The use of blockchain in the Satellite network, in particular, will allow the network to track and govern resource utilization and sharing more efficiently. Massive MIMO is a new technology for improving network capacity in multi-user scenarios. Paper has used a multi-user massive MIMO scenario with a spoofing attack. It examines the effects of eavesdropper spoofing pilot power and a variety of groups on the confidentiality efficacy of the systems under consideration.

Keywords Blockchain · Security · Massive MIMO · Bitcoin

1 Introduction

The 5G cellular network is a committed network that will give users high-quality coverage in a variety of industries including health banking and education. When relying on this technology, the most important thing to remember is security. To the subscriber, while the network will trust each other and also will share a symmetric key, the registration, authentication, as well as key agreement procedures are very important protocols in all cellular networks. A Third Generation Partnership Project set security criteria with the supplier for user authentication (3GPP). However, multiple studies focused on these protocols and determined that security issues still exist, proposing authentication and key agreement procedures as a result.

Blockchain is regarded as one of the evolving technologies which might get a significant impact on our lives in the next years. Blockchain is used by Bitcoin and smart contracts, for example, to provide security qualities like authenticity and validity to its applications [1]. To subtend the demands of new services also applications for the Internet of all, like multi-gigabit transfer rates, higher efficiency, sub-1 ms latency, and ubiquitous connectivity, the Satellite network must outperform previous generations (IoE). Although, given the scarcity of spectrum resources, good resource

A. Mohammed · S. R. Chopra (✉)
Lovely Professional University, Phagwara, Punjab, India
e-mail: shakti.chopra@lpu.co.in

D. Singh et al. (eds.), *Computer Aided Constellation Management and Communication Satellites*, Lecture Notes in Electrical Engineering 987,
https://doi.org/10.1007/978-981-19-8555-3_7

management also sharing are critical to accomplishing all of these high goals. The blockchain is one possible technique for doing all of this. Due to its fundamental qualities, which are critical to the Satellite network also other networks, the blockchain has lately attracted attention. The adoption of blockchain in Satellite, in particular, will allow the network to more efficiently track and control resource usage as well as sharing.

As a result, we'll look at how blockchain might aid in resource management and sharing in Satellite through a variety of scenarios such as the IoT, device-2-device communications also network slicing, as well as inter-domain blockchain ecosystems [2]. With its transparency, decentralization, also security properties, blockchain is a groundbreaking technology that is having a significant impact on modern society. Blockchain technology is poised to change the method we work, communicate, as well as to conduct business shortly. Academics, business leaders, also researchers are studying various facets of Blockchain as a new technology [3]. An unidentified individual (or people's group) named Satoshi Nakamoto initially presented the concept of Bitcoin. Bitcoin is a p2p cryptocurrency and a decentralized worldwide payment method for digital money which enables subscribers to make transactions outwardly without the necessity of a middleman.

Bitcoin transactions are conducted as well as verified by network nodes before being published in a public ledger known as the blockchain, which is administered through network entities working with Bitcoin software [4]. Blockchain protects in particular by authenticating peers who exchange virtual currency, encrypting data, as well as generating hash values [5, 6]. Blockchain is a decentralized technology. When it comes to solving corporate difficulties, it wields considerable power. The records of a blockchain transaction are encrypted, and each transaction is connected to the previous transactions or records. Blockchain dealings are validated by algorithms running on nodes. A single individual cannot begin a transaction. Finally, blockchains promote transparency by letting all participants see all transactions at any moment. Blockchain's fundamental properties are decentralization and immutability. Deals are finished more quickly; transactions and certification are accomplished in seconds, and so forth [7].

Massive MIMO is a prospective technology for next-generation wireless communication networks. Its application to ensure communication protection has piqued the interest of many. Hundreds of antennas have been installed at the base station in massive MIMO; also, the ensuing spatial-wideband impact has been studied. The effect of the large MIMO alternation for example on secrecy was explored [8]. Modern networks confront enormous traffic needs as a result of globalization, and to meet these expectations, cellular systems are installed within a little hundred meters of one another; also, wireless LANs are established practically anywhere. Wireless traffic is being boosted by the advent of new technologies such as the IoT and machine-2-machine communication, as well as enhanced mobile broadband coverage.

The multiple input multiple output systems are an important part of today's wireless networks; also, they've been increasingly popular in recent years as a means of achieving excellent spectral and energy efficiency. SISO systems were popular before

MIMO, but they had limited throughput and couldn't manage a large number of users well [9]. New MIMO technologies, including SU-MIMO, MU-MIMO, and network MIMO, have been developed to accommodate this tremendous user demand. These technological developments, however, are insufficient to meet ever-increasing needs. In the last years, the number of wireless users has exploded, and these subscribers create trillions of bytes of data that have to handle efficiently and safely. The most appealing wireless connectivity infrastructure for five generations and beyond is massive MIMO. Massive MIMO in wireless networks is a move away from current MIMO systems, integrating 100 s or even 1000 s of antennas at the BS to accommodate tens of users at once. In theoretical channels, discrete freelance and identically distributed complex Gaussian coefficients are separated.

The majority of research has so far been on Rayleigh channels. This research investigates the performance of large MIMO in channels that have been measured in real-world spread settings. A massive MIMO without cells is a possible future wireless networking approach in which several geographically distant access points which work together to serve individual subscribers in the same frequency–time resource. Cell less massive MIMO will provide superior coverage, spectral performance, and energy efficiency than both tiny cells and co-located massive MIMO. In cell-free large MIMO networks, performance computation, power control, and beamforming have all been investigated further. A user-centric strategy was also studied, as well as a downstream pilot in cell-free massive MIMO. Using a pilot spoofing attack, the authors deliberate the security of the cell-free massive MIMO system. It also included optimal power allocation strategies for achieving the highest data rate or secrecy rate possible [10].

Massive MIMO is a relatively new wireless networking architecture that has gotten much buzz in latest years. Multi-user MIMO systems with a large number of antennas are referred to as huge MIMO (10–100 s). The base station may only have up to 8 antennas in comparison with the LTE standard. Massive MIMO thus scales traditional MIMO by a factor of two. Frequency resource, a base station with a no of antennas frequently supports a lot of single-antenna customers simultaneously. Theoretically, such systems have the potential to increase connection reliability, and spectral quality, as well as transmit energy efficiency significantly.

The sufficiently complex scattering environment can be interpreted as "favorable" propagation. In these circumstances, even basic linear precoding/detection systems like zero forcing as well as matched filtering become almost ideal. Massive MIMO's enticing qualities, on the other hand, are centered on optimistic propagation assumptions mixed with low-cost technology that enables the deployment of a large number of antennas. Most research has thus far concentrated on theoretically independent and identically distributed complicated Gaussian that is Rayleigh fading, channels with unbounded antenna numbers [11]. The spatial focusing can be sharper with more antennas. The canonical massive MIMO system employs the time division duplex mode, in which the uplink, as well as downlink broadcasts, share the very same frequency resource and are timed independently.

Physical propagation channels will be reciprocal, which means that their response is the same in both directions, and thus, they may be utilized in a time division duplex.

Massive MIMO systems, in particular, leverage reciprocity to forecast uplink channel responses before employing the CSI for uplink receive combining and downstream payload data send precoding. Because most transceiver hardware isn't reciprocal, channel reciprocity must be fine-tuned to reap the benefits in practice. Providentially, the uplink–downlink hardware mismatches only change a little degree in an hour, and they may be reduced with easy relative calibration ways that don't need the installation of extra reference transceivers and rely merely on the mutual coupling between array antennas. [12]. A massive MIMO is a very recent 5G wireless access solution that works in both mm-Wave and sub-6 GHz bands. Since its inception around a decade ago, it has progressed from a bizarre "academic" notion to the core technology that will very certainly be employed in all future wireless technologies. We can say massive is an MU-MIMO technology that can deliver wireless terminals in high-mobility scenarios with consistently acceptable service. The basic concept is to provide BSs with numerous antenna arrays that may serve several terminals on the same time–frequency resource [13].

The base station array learns the channel in both directions using channel estimations received from uplink pilots transmitted by the terminals. Massive MIMO base stations are self-contained, with no data or channel state information shared with neighboring cells. As a result, massive MIMO has been completely scalable in terms of the number of BS antennas. A massive MIMO is a relatively new wireless networking architecture is generated a lot of buzzes these years. Huge MIMO refers to MU-MIMO systems with a huge number of antennas on the BS (say, 10–100 s). In comparison, the LTE standard only supports up to eight antennas at the BS [14].

2 List of Top Blockchain Features

2.1 Immunity

There are several exciting blockchain properties, but "immutability" is without a doubt the most important. But how come this technology is still not tainted? Let us just start with a connecting blockchain that is immutable.

The inability to change or modify anything is referred to as immutability. This is among the most crucial blockchain elements for maintaining the technology's current state—a stable, unchanging network. So how can you make sure it stays that way? In a little respect, blockchain technology differs from conventional financial systems. Rather than depending on centralized authority, it includes the blockchain features throughout a network of nodes.

2.2 Decentralized

As we know the network is decentralized, which implies the infrastructure is not controlled by a single entity or individual. This network is currently decentralized, as it is managed by a community of nodes. One of the most useful characteristics of blockchain technology is this. Let me try to make things easier for us. The blockchain assigns us, the customers, and a straightforward job. As it will not require any regulating body, we may access the framework straight from the web and also keep our resources there. Cryptocurrencies, important documents, contracts, and other essential digital assets may all be saved. And, due to blockchain, you'll become able to use your private key to assert complete control over them. As a result, the decentralized system restores the power and privileges of ordinary people on their property.

2.3 Enhanced Security

Because no one can readily change the network's properties for their benefit, so will be no need for a central authority. Encryption provides an additional degree of security to the device.

When cryptography is paired with decentralization, consumers gain an additional degree of protection. Cryptography is a difficult mathematical procedure that prevents data from being intercepted. On the blockchain, every data is cryptographically coded. To put it another way, network information hides the underlying properties of the data. Each incoming data is run through a mathematical procedure that creates a new value; however, the length stays unchanged. You can think about it as a type of record identifier. Every block in the ledger does have its hash, also the last block's hash. Every hash ID might be updated if the data was changed or attempted to be modified. And even that is a stretch.

2.4 Distributed Ledgers

In most cases, a public ledger can hold the necessary information about a transaction as well as its users. On the contrary, the justification for private or federated blockchain is different. In such cases, however, a significant number of individuals will be able to observe what is truly going on in the ledger.

Hence, it has been regarded among the most significant elements of the blockchain. So, the ultimate result is a more efficient ledger system which might be competing with traditional ledger systems.

2.5 Consensus

Each blockchain relies on consensus algorithms to function. The architecture, which is carefully created, is built around consensus methods. A consensus mechanism is built into every blockchain to help the network make choices. Consensus is a decision-making process for the network's active nodes in its most basic form. In this instance, the nodes should be able to reach an agreement fast and effortlessly. While millions of nodes are validating a transaction, a system's ability to run smoothly depends entirely on consensus. The consensus is to blame for the lack of trust in the network. So the algorithms at the center of the system can be trusted; however, nodes cannot trust each other. As a result, any network move that benefits the blockchain is a win–win situation. One of the benefits of blockchain technology is this.

Around the world, numerous blockchain consensus algorithms are in use. Each has a unique way of making judgments, and perfecting prior approaches generate faults. The design of the Internet establishes a sphere of justice. However, every blockchain should provide a consensus mechanism to sustain decentralization; furthermore, the blockchain's essential value would be lost.

2.6 Faster Settlement

Traditional financial networks are slow and inefficient. Processing a contract will take many days after all settlements have been finalized. It's also simple to mess with. Blockchain enables faster settlement as compared to traditional financial systems. It enables a subscriber to transmit money very quickly; these blockchain attributes are extremely beneficial to multinational personnel and explain.

They may now send money to their loved ones swiftly since blockchains are now much too fast for them. Aside from that, there's the smart contract system. It can help to speed up the resolution of any contract. This is still one of the very important blockchain features today. Now that the third party has been removed, anyone can send money for a minimal cost. So, what's stopping you from using blockchain technology? Although there are several instances where the network cannot accommodate a large number of users and a quick settlement is not feasible. Nonetheless, many are working to improve this situation, and we will soon see a better solution [15].

3 Benefits of Blockchain Security

3.1 Blockchain Basis

In the cryptocurrency and ledger-keeping industries, blockchain is critical. Policymakers, mobile operators, and even infrastructure commissioners have all taken notice of the technology because of the community's vibrancy. Blockchains are decentralized databases that are structured to use a hash tree2, which would be tamper-proof and irreversible by nature. It can add distributed trust and is also designed to provide database transaction coherence. Additionally, atomicity, sturdiness, accuracy, and data integrity are all possible with the blockchain.

Aside from the nature of its chain-link data structure, blockchains rely heavily on the consensus mechanism that provides an apparent ordering of transactions as well as the blockchain's integrity and coherence among geographically spread nodes. The CM is mostly responsible for the blockchain system's performance, including transaction throughput, latency, node scalability, and security level, among other things. As a result, multiple CMs might be chosen depending on application situations and performance needs. Practical Byzantine Fault Tolerance, PoW, and PoS are examples of commonly used CMs, and extensive studies of their performance and security, as well as how they may be utilized in various resource management also sharing situations, are described in Section.

3.2 Consensus Effects and Security Performance

If the blockchain's amazing also resistant data structure is the exterior of a house, the consensus is the foundations. On the CM, blockchain has several choices. The most important step in creating a stable and effective blockchain system is selecting a suitable consensus for Satellite resource management. CM is essential to blockchains because it dictates their efficiency in terms of TPS, latency, node scalability, protection, and so on. It assures the blockchain's integrity and consistency among geographically scattered nodes, as well as a clear ordering of transactions. The chain can be divided into two types based on access criteria: public and private. The public chain is permission less, relying on proof-of-concept consensus to offer a stable, dependable network for all users without demanding their identities at access points. On an ad hoc basis, there are possible anonymous clients and suppliers in the 6G resource pool. The advantage of using a public chain in ad hoc networks, when identity and security barriers are torn down enabling panoptic information flows, is significant. As a result, public chains can improve group productivity and control participant order. The scheme, on the other hand, would be vulnerable to breaches and malicious activities if participants were hidden. The consortium of private chains, on the other hand, is authorized, implying that access is restricted. It has a very consistent group makeup, and the members' identities are not kept hidden. While the network is less vulnerable to unknown attacks, it still has issues, such as the malicious byzantine node. When it comes to implementing new technology, the first concerns are usually security and dependability. In terms of security efficiency and resilience, blockchain technology was intended to exceed current solutions [2]. The Bussgang theorem is used to create a DAC quantization model for calculating the asymptotic achievable secrecy rate; the DAC decision does not influence the secrecy rate because the detrimental effect of quantization noise may be compensated for by decreasing the AN power [16].

4 Active Spoofing Attack on Cell-Free Massive MIMO Networking

4.1 Uplink Transmission

The property of channel reciprocity is employed in massive MIMO. Different groups are assigned to different pairwise orthogonal pilot sequences of length τ_p.

We consider a multi-group multicast cell-free massive MIMO system with M single antenna APs in the presence of an active eavesdropper. Furthermore, each cell includes T groups of users with L single antennas.

The nth group of users provided the pilot sequence. Because the pilot sequences are public and standardized, an eavesdropper can easily spoof the legitimate user's pilot sequences to change the beam direction. We can suppose that the purpose of the eavesdropper is to intercept confidential messages intended for the nth group while maintaining generality.

$$\sqrt{\tau_p \rho_E} = \sqrt{\tau_p \rho_k}$$

where $\sqrt{\tau_p \rho_k}$ denotes the pilot sequence of an eavesdropper, f_{nt} and f_{nE} are the channel coefficients, and N_0 additive white Gaussian noise vector, As a result, the received pilot vector at the nth AP is

$$y_n^p = \sqrt{\tau_p \rho_k} \sum_{t=1}^{T} \sum_{l=1}^{L} f_{nt} \varphi_t + \sqrt{\tau_p \rho_E} f_{nE} \varphi_t + N_0 \quad (1)$$

where $\rho_k(\rho_E)$ is the user's normalized average signal to noise ratio. N_0 is the additive white Gaussian noise vector with N_0 ~ CT (0, I). By projecting y_n^p into φ_t

$$y_{nt}^{(p)} = \sqrt{\tau_p \rho_k} \sum_{l=1}^{L} f_{nt} + \sqrt{\tau_p \rho_E} f_{nE} \delta(t' - t) + \varphi_{t'}^K N_0 \quad (2)$$

where $\delta(0) = 1$ and $\delta(t') = 0$ and $t' \neq 0$ for the channels estimated using the least mean squared error method.

4.2 Downlink Multicasting Transmission

At the nth AP, the transmit multicasting signal might be provided by

$$R_n = \sqrt{\frac{\rho_d}{T}} \sum_{t=1}^{T} \frac{Z_{nt}^*}{|Z_{nt}|} S_t \tag{3}$$

where $Z_{nt} \sim \text{CT}\ (0, 1)$

$$\{ y_{t_l} = \sqrt{\rho_d} b_{tl} S_t + \sqrt{\rho_d} \sum_{t' \neq t}^{T} b_{t_l'} S_{t'} + N_{0tl}$$

$$\{ y_E = \sqrt{\rho_d} b_t E S_t + \sqrt{\rho_d} \sum_{t' \neq t}^{T} b_{t'} S_{t'} + N_0, \tag{4}$$

where $b_{t_l} \triangleq \sum_{m=1}^{M} \sqrt{\frac{1}{L}} \left(\frac{(f_{nt_l} Z_{nt}^*)}{(|Z_{nt}|)} \right)$, $b_{t_l'} \triangleq \sum_{m=1}^{M} \sqrt{\frac{1}{L}} \left(\frac{(f_{nt_l} Z_{nt'}^*)}{(|Z_{nt'}|)} \right)$, $b_{tE} \triangleq \sum_{m=1}^{M} \sqrt{\frac{1}{L}} \left(\frac{(f_{nE} Z_{nt}^*)}{(|Z_{nt}|)} \right)$, $b_{t'E} \triangleq \sum_{m=1}^{M} \sqrt{\frac{1}{L}} \left(\frac{(f_{nE} Z_{nt'}^*)}{(|Z_{nt'}|)} \right)$, where ρ_d is the downlink transmission normalized SNR.

5 Result and Discussion

Results derived by applying uplink and downlink using different pair of antennas; it is found that the use of twenty pair of antenna provides the better result as compared to single and five pair as shown in Fig. 1.

6 Conclusion and Future Scope

In this paper, we have discussed how to secure the multi-layer blockchain and how to make it reliable in distribution. Modern networks confront enormous traffic needs as a result of globalization, and to meet these expectations, cellular systems are installed within a little hundred meters of one another; also, wireless LANs are established practically anywhere. Wireless traffic is being boosted by the advent of

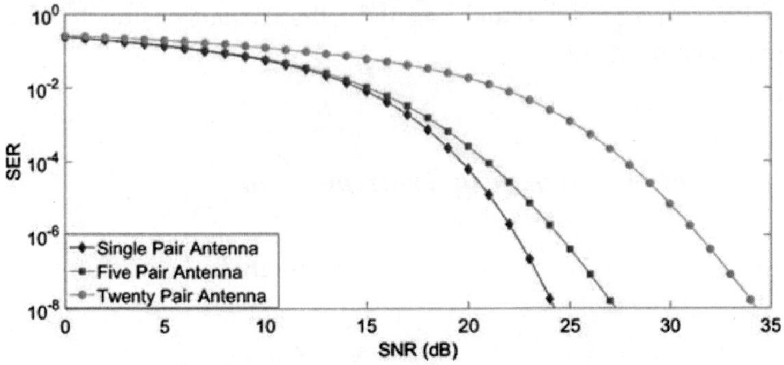

Fig. 1 SER versus SNR in massive MIMO scenario with a spoofing attack

new technologies such as the IoT and machine-2-machine communication, as well as enhanced mobile broadband coverage.

References

1. Haddad Z et al (2020) Blockchain-based authentication for 5G networks. In: 2020 IEEE international conference on informatics, IoT, and enabling technologies (ICIoT). IEEE, 2020
2. Xu H et al (2020) Blockchain-enabled resource management and sharing for 6G communications. Dig Commun Netw 6(3):261–269
3. Hussein DM, Taha MH, Khalifa NE (2018) A blockchain technology evolution between business process management (BPM) and Internet-of-Things (IoT). Int J Adv Comput Sci Appl 9(8):442–450
4. Rahouti M, Xiong K, Ghani N (2018) Bitcoin concepts, threats, and machine-learning security solutions. IEEE Access 6:67189–67205
5. Park JH, Park JH (2017) Blockchain security in cloud computing: use cases, challenges, and solutions. Symmetry 9(8):164
6. Dasgupta D, Shrein JM, Gupta KD (2019) A survey of blockchain from a security perspective. J Banking Fin Technol 3(1):1–17
7. Stephen R, Alex A (2018) A review on blockchain security. IOP Conf Ser: Mater Sci Eng 396(1)
8. Chataut R, Akl R (2020) Massive MIMO systems for 5G and beyond networks—overview, recent trends, challenges, and future research direction. Sensors 20(10):2753
9. Zhang X, Guo D, An K (2019) Secure communication in multigroup multicasting cell-free massive MIMO networks with active spoofing attack. Electron Lett 55(2):96–98
10. Gao X et al (2015) Massive MIMO performance evaluation based on measured propagation data. IEEE Trans Wirel Commun 14(7):3899–3911
11. Björnson E, Hoydis J, Sanguinetti L (2017) Massive MIMO has unlimited capacity. IEEE Trans Wireless Commun 17(1):574–590
12. Björnson E, Larsson EG, Marzetta TL (2016) Massive MIMO: ten myths and one critical question. IEEE Commun Mag 54(2):114–123
13. Larsson EG et al (2014) Massive MIMO for next-generation wireless systems. IEEE Commun Mag 52(2):186–195
14. Chopra SR, Gupta A (2021) Error analysis of grouped multilevel space-time trellis coding with the combined application of massive MIMO and cognitive radio. Wirel Pers Commun 117(2):461–482

15. Chopra SR, Kaur N, Monga H (2016) Space-time coding techniques in MIMO: a review. Indian J Sci Technol 9(47):1–5
16. Xu J et al (2019) Secure massive MIMO communication with low-resolution DACs. IEEE Trans Commun 67(5):3265–3278

Design of Small SAR Constellation for Minimizing Revisit Time

Vetal Akshay Pandit, Ameya A. Kesarkar, Yogendra Sahu, Ashok Rohada, J. Rao, Pankaj K. Nath, Rakesh Bhan, Ch. V. N. Rao, and Rajeev Jyoti

Abstract In this paper, we propose a novel formulation to obtain the small SAR constellation that optimizes revisit times in the worst, average, and best scenario over a given Region of Concern (ROC). We demonstrate this formulation for selected Indian regions via detailed System Tool Kit (STK) simulations to get the optimum constellations.

Keywords Constellation · Small SAR · Revisit time · Optimization

1 Introduction

When multiple satellites are put in orbits in such a way that their coordination can fulfil the objective of one giant satellite, they are called 'formation flying satellites'. In general, we have three possible flying formations: (1) trailing, (2) clustering, and (3) constellation. Trailing formation is formed by multiple satellites orbiting the same path. Each one follows the previous one separated by a specific time interval. Trailing satellites are especially suited for meteorological, environmental, and surveillance of specific target applications. Cluster is formed by satellites in dense arrangement, such as twin satellites. Constellation can be considered as a number of satellites with coordinated ground coverage, operating together under shared control. Usually, it is made up of small satellites. There are various classical constellation design methods including Walker [1], Rosette [2], and tetrahedron elliptical constellation [3], which follow geometric design approach for uniform symmetric constellation pattern [4].

Constellation of SAR satellites in Low Earth Orbit (LEO) is a relatively new concept. ICEYE—Finland and Capella Space—US started a revolution in SAR imaging using constellation [5–7]. They are designed to provide persistent monitoring capabilities and high-resolution view of the Earth's surface. Primary advantage of constellation is smaller revisit time [8–12]. The work in [13] proposes a novel technique to compute the revisit time of satellites within repeat ground tracks.

V. Akshay Pandit (✉) · A. A. Kesarkar · Y. Sahu · A. Rohada · J. Rao · P. K. Nath · R. Bhan · Ch. V. N. Rao · R. Jyoti
Space Applications Centre (SAC), ISRO, Ahmedabad 380015, India
e-mail: vetalakshay@sac.isro.gov.in

D. Singh et al. (eds.), *Computer Aided Constellation Management and Communication Satellites*, Lecture Notes in Electrical Engineering 987,
https://doi.org/10.1007/978-981-19-8555-3_8

In [14], a method is given to address a complex coverage area by using sparse satellite constellation design with a limited coverage capability. For a given ROC, finding constellation configuration that optimizes revisit time, however, hasn't received much attention in the available literature.

In this paper, we focus on formulating a novel optimization problem to obtain best possible revisit of small SAR constellation over a given ROC. We also illustrate our approach via STK simulations.

2 Formulation of Optimization Problem for Constellation Design

Revisit time is the time elapsed between successive satellite passes over a given region. In Fig. 1, t_3 denotes the worst revisit time, which is the maximum of [t_1, t_2, t_3, t_4].

Figure 2 shows the general sketch of revisit time versus number of satellites profile. At one particular point even if we increase number of satellites further, revisit time does not show any significant improvement. This point shows the optimum number of satellites required for that region.

2.1 A Novel Framework for Worst, Average, and Best Revisit Optimization

For this purpose, ROC is considered as a set of equi-spaced grid points. In the worst revisit optimization case, we proceed as follows:

- For given no. of satellites per plane (m) and number of orbital planes (n), maximum revisit time is calculated for each grid point of ROC.

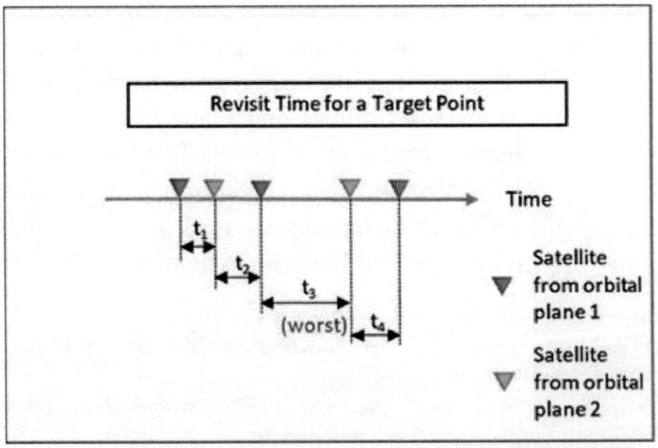

Fig. 1 Concept of revisit time

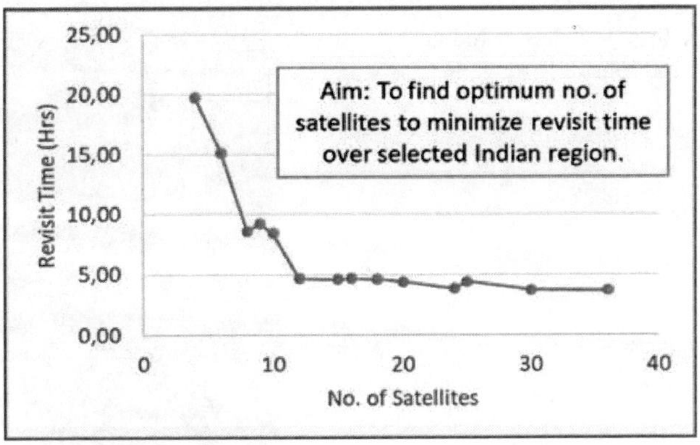

Fig. 2 Variation of revisit time against no. of satellites

- From ROC perspective, maximum of all grid point maxima is the worst scenario for given (m, n).

We would like to minimize this maximum w.r.t. total number of satellites (i.e. $m \times n$). Therefore, our novel optimization problem becomes nothing but the minimization of such ROC maxima w.r.t. $m \times n$ satellites as presented below:

$$\text{Worst Revisit} = \min\left(\max\left(\max\left(\begin{array}{c}\text{revisit time}\\ \text{for each grid}\end{array}\right)\right)\right)$$

In the similar manner, Average and Best revisit times can be calculated based on the frameworks as proposed below:

$$\text{Average Revisit} = \min\left(\max\left(\text{avg}\left(\begin{array}{c}\text{revisit time}\\ \text{for each grid}\end{array}\right)\right)\right)$$

$$\text{Best Revisit} = \min\left(\max\left(\min\left(\begin{array}{c}\text{revisit time}\\ \text{for each grid}\end{array}\right)\right)\right)$$

3 Small SAR Constellation Over Selected Indian Regions

Figure 3 shows ROCs for simulation purpose which are as follows: (1) Whole India and (2) North Border of India.

Minimizing Revisit Time. First, the optimum combination of number of satellites and number of planes for any region needs to be identified. For given number of satellites and number of planes, we select Walker-Delta constellation [1]. All orbital planes are assumed to be in same inclination with reference to the equator. We use Walker-Delta Constellation Tool of STK for our simulation purpose.

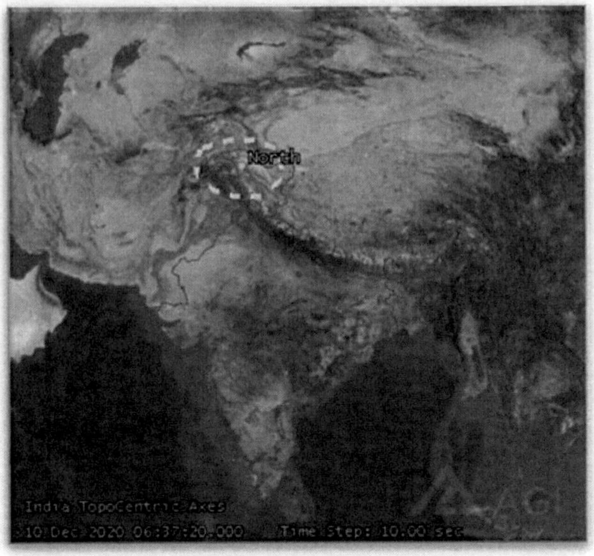

Fig. 3 Selected Indian regions

In STK, any region is considered as set of grids separated with grid spacing angle. Finer the spacing angle more accurate will be the results. In our design, grid spacing (latitude/longitude) is placed at 0.5°. Analyzer tool in STK gives provision to optimize the required parameters with the help of Parametric Study Tool and Carpet Plot Tool.

Considering the Indian region's latitude range, we begin with orbital plane inclination angle of 40°. With SAR altitude of 560 km, other orbital parameters of first satellite are given in Table 1.

A conical beam sensor is added in our satellite. Right as well as left look of sensors are considered as satellite which has flexibility to orient its axis both direction mechanically. Figure 4 shows the left and right look beam of the satellite.

In this case, no. of satellites per plane and no. of planes are varied from 2 to 5, which means that the total no. of satellites vary from 4 to 25. Average, worst and best

Table 1 Orbital parameters of first satellite

Parameter	Value
Semi-major axis	6938.14 km
Eccentricity	3.21947e−13
Inclination	40°
Argument of Perigee	0°
RAAN	0° (equally spaced for different orbits as per walker constellation)
True anomaly	0° (equally spaced for different orbits as per walker constellation)

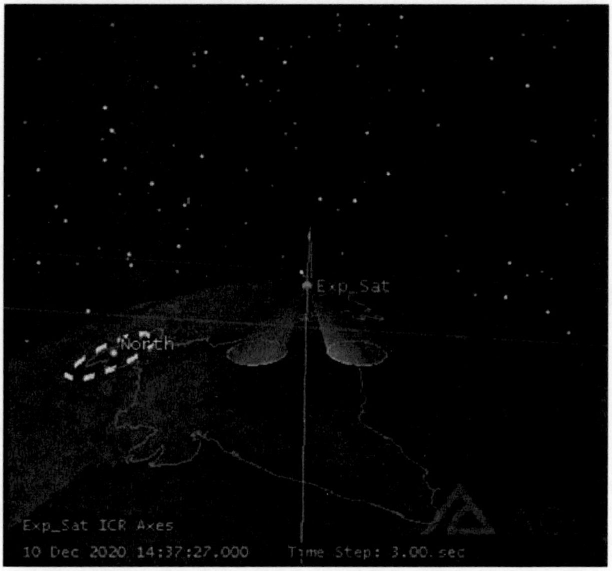

Fig. 4 Left and right look beams of satellite

revisit times are calculated in each case. On plotting revisit time against the number of satellites, we get the optimum configuration (as previously discussed in Sect. 2).

Region1: Whole India

Figure 5 shows STK view of constellation over whole Indian region.

Table 2 shows the average, worst, and best revisit times for different combinations of no. of planes and no. of satellites per plane by maintaining inclination angle at 40°.

Fig. 5 Constellation over Indian region

Table 2 Revisit time over Indian region for inclination angle of 40°

No. of planes		2	2	2	2	3	3	3	3	4	4	4	4	5	5	5	5
No. of satellites per plane		2	3	4	5	2	3	4	5	2	3	4	5	2	3	4	5
Revisit time (Hrs.)	Average	>24	4.8	2.7	2.6	>24	2.7	1.8	1.6	3.4	2.7	1.4	1.1	12.0	1.7	1.1	0.9
	Worst	>24	10.8	9.8	9.5	>24	7.9	7.0	6.8	10.7	4.9	4.6	4.4	22.4	4.7	3.7	4.1
	Best	>24	2.0	0.8	0.7	>24	0.5	0.8	0.3	0.8	1.1	0.8	0.3	19.4	1.1	0.4	0.3

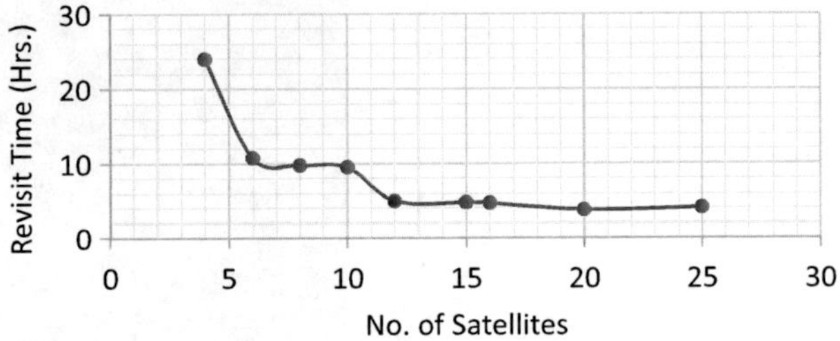

Fig. 6 Revisit vs no. of satellites for Indian region

We arrive at no. of planes = 3, no. of satellites per plane = 4 (i.e. total 12 satellites) as the optimum requirement in this case. The point of optimum can be seen as a red dot in Fig. 6 correspondingly.

In the next step, we fix the optimum configuration (3,4) and vary orbital inclination angle from 30° to 45° in the step of 1° to check if there is any further possible improvement in revisit time with respect to inclination angle. Consequently, we reach the inclination angle of 32°, as per the details presented in Table 3.

Region 2: North Border of India

Figure 7 shows STK view of constellation over North Border region. Table 4 presents that the optimum configuration of no. of planes and no. of satellites per plane is (3,4) at 40°. inclination angle. Figure 8 shows the corresponding plot of revisit time variation against total no. of satellites. By fixing this optimum (i.e. 3,4), when we vary inclination angle around 40°, we find that in this particular case, 40°. itself is the best possible inclination angle as given in Table 5.

Overall results are summarized in Table 6.

Table 3 Revisit time over Indian region for optimum configuration

Inclination (°)		30	31	32	33	34	35	36	37	38	39	40	41	42	43	44	45
Revisit time (h)	Average	>24	12.0	1.6	1.8	1.8	1.8	1.8	1.8	1.8	1.8	1.8	1.8	2.0	2.0	2.2	2.2
	Worst	>24	20.7	6.6	6.6	6.6	7.0	7.0	7.0	7.0	7.0	7.0	7.0	7.1	7.1	7.1	7.1
	Best	>24	0.4	0.4	0.8	0.8	0.8	0.8	0.8	0.8	0.8	0.8	0.8	0.8	0.8	0.8	0.8

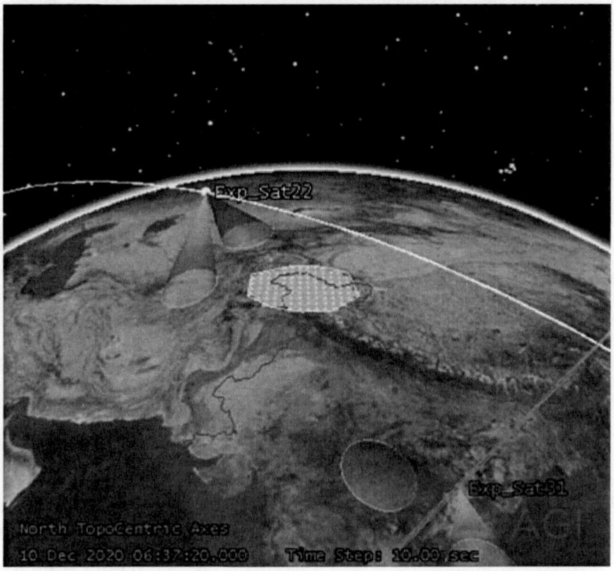

Fig. 7 Constellation over North Border region

Table 4 Revisit time over North Border region for inclination angle of 40°

No. of planes		2	2	2	2	3	3	3	3	4	4	4	4	5	5	5	5
No. of satellites per plane		2	3	4	5	2	3	4	5	2	3	4	5	2	3	4	5
Revisit time (Hrs.)	Average	2.7	1.8	1.4	1.1	1.7	1.2	0.9	0.7	1.3	0.9	0.7	0.6	1.1	0.7	0.5	0.4
	Worst	7.7	6.6	6.5	6.4	4.6	2.9	2.9	2.7	3.4	2.8	2.9	2.7	3.1	2.8	2.9	2.7
	Best	0.8	0.5	0.4	0.3	0.8	0.5	0.4	0.3	0.5	0.5	0.4	0.3	0.6	0.3	0.2	0.1

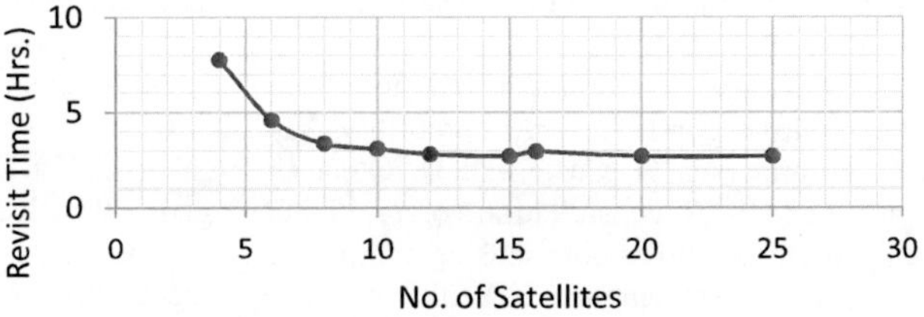

Fig. 8 Revisit vs no. of satellites for North Border region

Table 5 Revisit time over North Border region for optimum configuration

Inclination (°)		30	31	32	33	34	35	36	37	38	39	40	41	42	43	44	45
Revisit time (Hrs.)	Average	>24	>24	3.4	1.4	1.5	1.6	1.7	1.7	1.7	1.2	0.9	1.0	1.1	1.2	1.2	1.3
	Worst	>24	>24	7.5	5.9	5.4	4.6	4.2	3.8	3.8	2.9	2.9	3.4	4.2	4.6	5.1	5.1
	Best	>24	>24	0.4	0.4	0.4	0.4	0.4	0.4	0.4	0.4	0.4	0.4	0.4	0.4	0.4	0.4

Table 6 Overall results

		India	North Border
No. of planes		3	3
No. of satellites per plane		4	4
Inclination (°)		32	40
Revisit time (h)	Average	1.6	0.9
	Worst	6.6	2.9
	Best	0.4	0.4

4 Conclusion

In this paper, we covered our novel framework of small SAR constellation design to achieve optimum revisit times over the given ROC in the worst, average, and best-case scenario. For the demonstration purpose, we first considered whole India as ROC and obtained required number of satellites and corresponding revisit times via STK simulations for these three scenarios. The exercise was repeated for North Border region as ROC in the similar manner. Our results suggest that the average (i.e. typical) scenario it is possible to revisit whole India in 1.6 h and North Border in 0.9 h with 12 numbers of satellites.

References

1. Walker JG (1984) Satellite constellations. J Br Interplanet Soc 37:559
2. Ballard AH (1980) Rosette constellations of Earth satellites. IEEE Trans Aerosp Electron Syst AES-16(5):656–673
3. Draim JE (1987) A common-period four-satellite continuous global coverage constellation. J Guid Control Dyn 10(5):492–499
4. Sah R, Srivastava R, Das K (2021) Constellation design of remote sensing small satellite for infrastructure monitoring in India. arXiv preprint arXiv:2107.09253
5. Stringham C, Farquharson G, Castelletti D, Quist E, Riggi L, Eddy D, Soenen S (2019) The Capella X-band SAR constellation for rapid imaging. In: IGARSS 2019–2019 IEEE international geoscience and remote sensing symposium. IEEE, pp 9248–9251
6. Farquharson G, Woods W, Stringham C, Sankarambadi N, Riggi L (2018) The Capella synthetic aperture radar constellation. In: EUSAR 2018; 12th European conference on synthetic aperture radar. VDE, pp 1–5
7. Ignatenko V, Laurila P, Radius A, Lamentowski L, Antropov O, Muff D (2020) ICEYE microsatellite SAR constellation status update: evaluation of first commercial imaging modes. In: IGARSS 2020–2020 IEEE international geoscience and remote sensing symposium. IEEE, pp 3581–3584
8. Kim Y, Kim M, Han B, Kim Y, Shin H (2017) Optimum design of an SAR satellite constellation considering the revisit time using a genetic algorithm. International Journal of Aeronautical and Space Sciences 18(2):334–343
9. Kim H, Chang YK (2015) Mission scheduling optimization of SAR satellite constellation for minimizing system response time. Aerosp Sci Technol 40:17–32
10. Lang TJ, Hanson JM (1984) Orbital constellations which minimize revisit time. Adv Astronaut Sci 54:1071–1086

11. Williams EA, Crossley WA, Lang TJ (2001) Average and maximum revisit time trade studies for satellite constellations using a multiobjective genetic algorithm. J Astronaut Sci 49(3):385–400
12. Moreira A, Krieger G (2003) Spaceborne synthetic aperture radar (SAR) systems: state of the art and future developments. In: 33rd European microwave conference proceedings (IEEE Cat. No. 03EX723C), vol 1. IEEE, pp 101–104
13. Luo X et al (2017) A novel technique to compute the revisit time of satellites and its application in remote sensing satellite optimization design. International Journal of Aerospace Engineering
14. Savitri T et al (2017) Satellite constellation orbit design optimization with combined genetic algorithm and semianalytical approach. International Journal of Aerospace Engineering

Array Antenna Design and Development for X-Band Applications

K. Malaisamy, Mohd. Wasim, P. Sivagamasundhari, G. Sivakannu, and V. Dinesh

Abstract This paper proposes an antenna array particularly intended for ocean radar systems. The array is made up of 16 identical elements. This antenna resonates at 10.10 GHz, and the bandwidth of antenna is 90 MHz. The elements are organized along the extended arms of a four-element one-dimensional arrangement and are feed in the middle by a number of perfect finishing touch microstrip lines. To depart from the established route, each patch element was cut with a slit to allow for horizontal polarization.

Keywords Microstrip antenna · X-band · Advanced Design System (ADS) · Array antenna · Radar

1 Introduction

An antenna is a radiating device that sends electromagnetic energy into space from a transmitting antenna. The antenna is a critical component in wireless communication systems. An antenna that is well designed can reduce system requirements while

K. Malaisamy (✉) · P. Sivagamasundhari · G. Sivakannu
Department of Electronics and Communication Engineering, Saranathan College of Engineering, Trichy 620012, Tamilnadu, India
e-mail: malaisamy-ece@saranathan.ac.in

P. Sivagamasundhari
e-mail: sivagamasundhari-ece@saranathan.ac.in

G. Sivakannu
e-mail: sivakannu-ece@saranathan.ac.in

Mohd. Wasim
School of Electronics and Electrical Engineering, Lovely Professional University, Phagwara, Punjab, India
e-mail: mohammad.wasim221@gmail.com

V. Dinesh
Department of Electronics and Communication Engineering, Easwari Engineering College, Chennai, India
e-mail: dinesh.v@eec.srmrmp.edu.in

D. Singh et al. (eds.), *Computer Aided Constellation Management and Communication Satellites*, Lecture Notes in Electrical Engineering 987,
https://doi.org/10.1007/978-981-19-8555-3_9

improving overall system performance. An antenna is utilized in broad range of applications, including land personal mobile phones, wireless personal area, and navigation equipment. As a result, demand for better capacity networks and compact smart phones that allow for more mobility is driving studies in this field. As a result of these factors, distinctive small and smart antennas must be considered as part of the solution. During the design stage of these antennas, designers must keep two things in mind: the physical and ergonomic qualities.

The patch antenna looks to be suitable for such applications, since it is lightweight, can be customized to adapt to its structural support and is economical when surplus. However, the patch antenna's greater physical characteristics are offset by its relatively poor electrical attributes. This antenna is frequently distinguished by a narrow bandwidth, low efficiency, and a high cross-polar level. In the case of arrays, surface waves can also have a negative impact on side lobes and directional gain. The antenna is also more expensive to build due to its limited bandwidth. The satellite communication antenna needs gain with minimum size with circular polarization. In order to increase the gain of the antenna array antenna, stacked array antennas are introduced.

Tomas Mikulasek et al. demonstrated a 22 patch antenna array for radar applications. The working frequency of this antenna is 24 GHz, and the return loss is −32 dB. The antenna's gain is 11.1 dBi [1]. Huayan Jin et al. proposed a wideband patch antenna array with a planar differentially L-shaped feed scheme. The proposed array was supplied using a planar, L-shaped feeding system to provide wideband functionality. The operating frequency ranges from 11.6 to 15 GHz, with a gain of 11.52 dBi [2]. Muhammad Ikram and colleagues created a simple single-layered continuous frequency and polarization-reconfigurable patch antenna array with a frequency band of 3.6 to 4 GHz. The 22 patch array was constructed and tested [3]. Andrew et al. created a perpendicular feed substrate circularly polarized stacked patch antenna. The frequency of operation for this antenna is 20 GHz. The reflection coefficient, gain, and radiation patterns of an eight-element linear array prototype are investigated [4]. Jun and colleagues developed a stacked wideband quad-polarization reconfigurable patch array antenna. The wideband antenna element is constructed on a double-layer structure that contains a driven substrate integrated waveguide (SIW) cavity and radiation patches. This antenna operates at 5.7 GHz. The antenna array's greatest gains have been recorded to be 9.85 dBic and 10.1 dBi [5].

Jin and colleagues demonstrated differentially fed patched multiple antennas with low cross-polarization and high bandwidths. This antenna has two operating frequencies: 12.62 GHz and 13 GHz, with gain values of 10.41 and 12.32 dBi, respectively, [6]. Yoo et al. proposed a twin connected feeding patch array antenna for 79 GHz automotive radar applications. A rectangular radiator and a waveguide feeding arrangement are proposed for the radar antenna. This antenna works at 79 GHz and has a maximum gain of 11.8 dBi [7]. Wide band, multiband, and stacked array antennas have been developed for a variety of purposes [8–13]. Cross-dipole array antennas and MIMO array antennas were developed for a variety of applications including radar, satellite, WLAN, Wi-Fi, and LTE [14–19].

According to the preceding literature review, antenna properties such as bandwidth, gain, and directivity are extremely low. There have been several efforts to realize the antenna for radar applications. This paper considers a 16-element microstrip array antenna for X-band applications. The study is organized as follows: Sect. 2 describes the geometry of the microstrip patch array antenna. Section 3 contains the conclusion and discussion. Section 4 concludes the paper.

2 Antenna Design

A 16-element array antenna is especially intended for maritime radar applications in the Advanced Design System (ADS) tool. The array is made up of 16 square elements that operate at a centre frequency of 10.10 GHz and a bandwidth of 90 MHz. The dimensions of the array are 11 cm × 4 cm × 1.8 mm. A single patch is 9.2 mm × 9.2 mm. To allow for horizontal polarization, a slit was bored within each patch element to change the flow of current. A vertical probe feeds the feeding network, which is made up of two *T* joints connected together. Every sequentially fed patch element is linked to the appropriate *T* junction arm.

The radiation parameters of an array antenna are examined for marine remote sensing applications. The array is composed of sixteen similar squared microstrip patches, each with four elements and arranged all along arms outstretched of a one-dimensional array. The amount of cross-polarization is greatly decreased due to the equal design in terms of the x-series. In this new design, each patch has a slit cut into it, and the number of patches is significantly decreased when compared to traditional patch antennas. The schematic structure of the antenna array designed in the Advanced System Design (ADS) tool is shown in Fig. 1. The antenna structure, seen in Fig. 1, consists of 16 patch components. Both top and bottom arrays are supplied out of phase currents due to the antenna array being placed head-to-head.

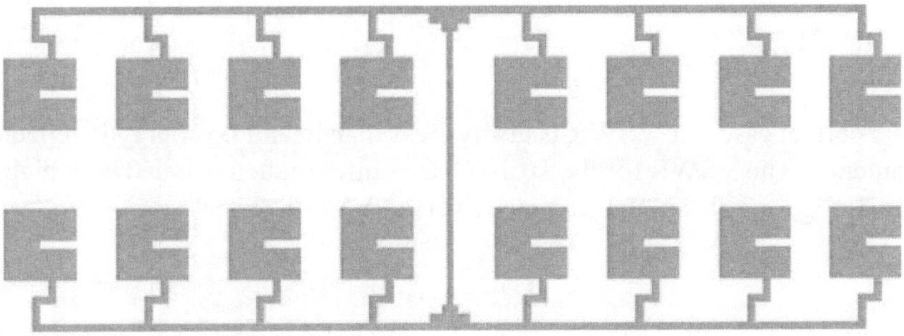

Fig. 1 Element antenna array structure

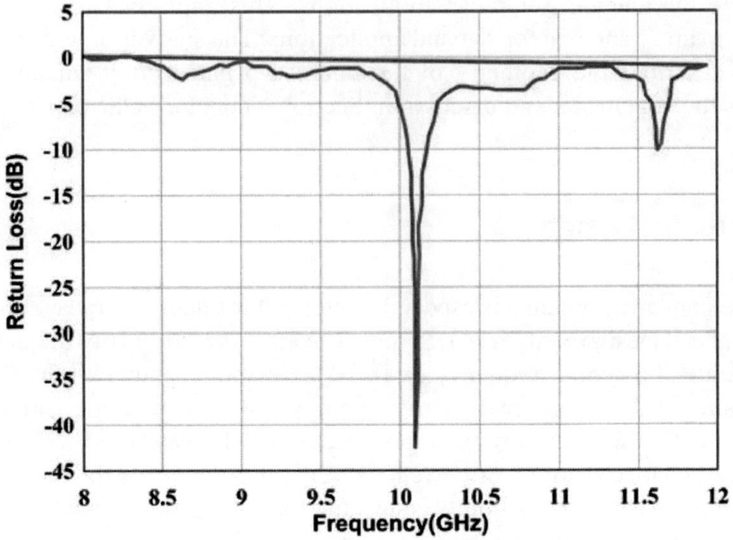

Fig. 2 Frequency vs return loss plot

3 Result and Discussion

3.1 Return Loss

For the proposed antenna, we produced a return loss of more than −41.931 dB at resonant frequency of 10.10 GHz. Figure 2 depicts a frequency versus S11 plot.

3.2 VSWR

In this perfect case, the VSWR is always less than 2, and no energy is reflected from the antenna. The VSWR for the 10.10 GHz centre frequency is 1.043, which is close to ideal. Figure 3 shows the frequency versus VSWR plot.

3.3 Radiation Pattern

A radiation pattern describes the change in power emitted by an antenna as a function of direction away. At 35° elevation angle, the maximum power is attained. Figure 4 depicts the designed antenna's radiation pattern.

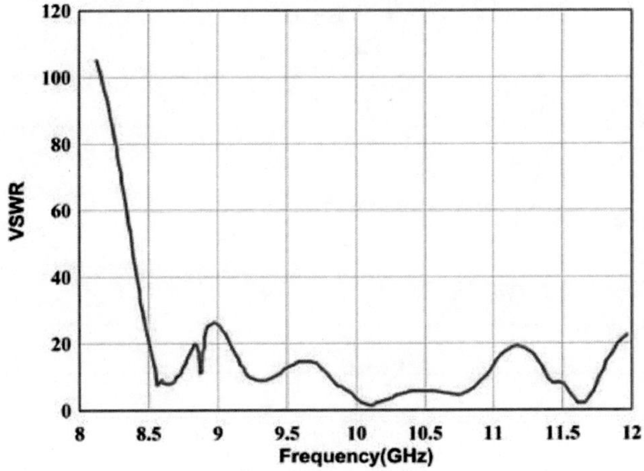

Fig. 3 Frequency versus VSWR plot

Fig. 4 Radiation pattern of the proposed antenna

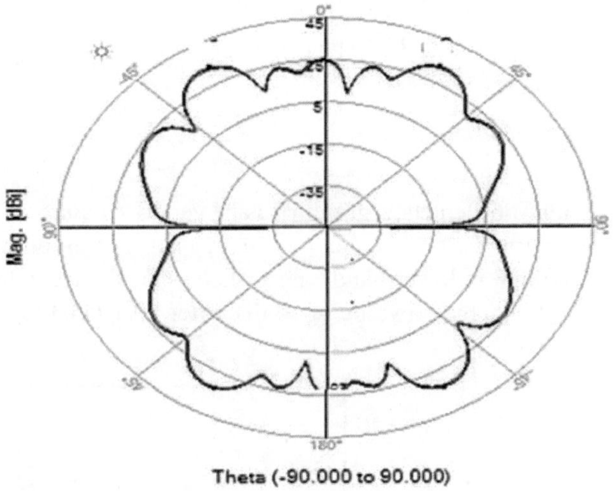

3.4 Circular Polarization

The circular polarization of an antenna is a polarization in which the electromagnetic current of the travelling wave changes direction in a circling pattern rather than changing intensity. Figure 5 depicts the simulated axial ratio result. This result reveals an excellent axial ratio less than 3-dB at the design frequency, indicating appropriate circular polarization.

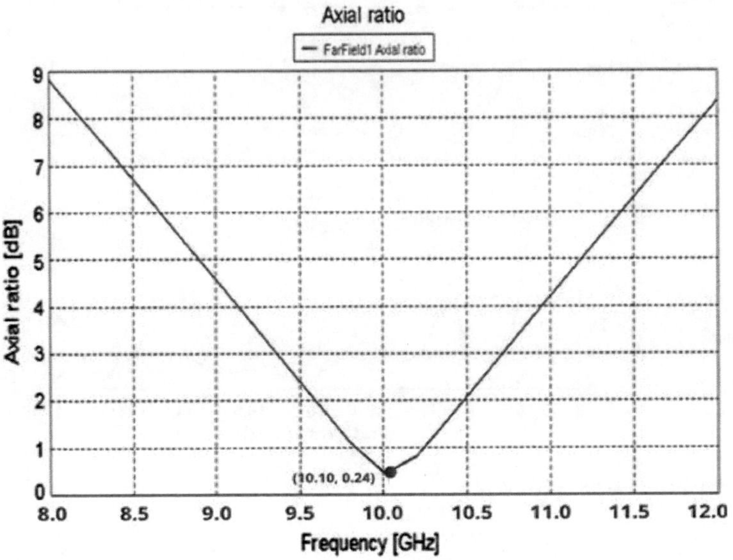

Fig. 5 Axial ratio of the proposed antenna array

3.5 Gain

Figure 6 depicts a graph of gain versus frequency. Figure 6 demonstrates a rising gain trend when frequency is increased. The amplitude of gain looks to be relatively minimal in lower bands; nevertheless, as the frequency increases, the gain grows as well, reaching a value of 14 dBi after 10.10 GHz.

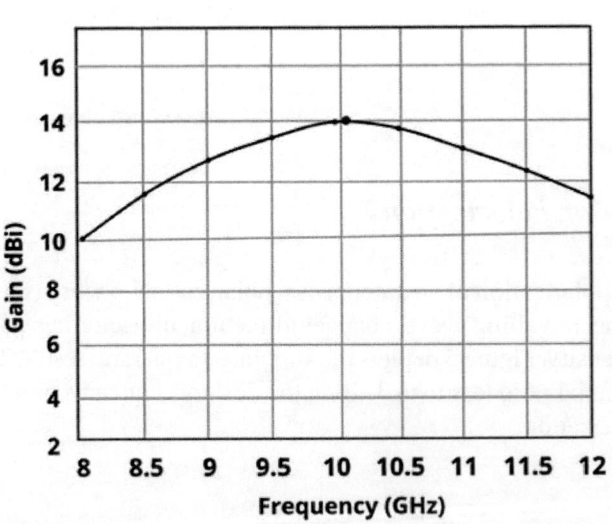

Fig. 6 Gain versus frequency plot

Table 1 Comparison between proposed and other microstrip patch array antennas

Reference	Frequency (GHz)	Return loss (dB)	Gain (dBi)
[1]	24	−22	11.1
[2]	11.6–15	−30	10.3–11.52
[3]	3.6–4	−25	5–10
[4]	20	−32	12.1
[5]	5.7	>−15	9.85
[6]	12.62, 13	>−15	10.41, 12.32
[7]	79	−35	11.8
Proposed work	10.10	−41.93	14

The suggested antenna's functional properties, such as frequency, return loss, and gain are compared to the reported antenna, which are given in Table 1. In the described articles, a trade-off between gain and return loss or size is necessary, which is relieved by the proposed antenna. The proposed antenna has a return loss of −41.93 dB. For a frequency of 10.10 GHz, the suggested antenna has a maximum gain of 14 dBi. Furthermore, the proposed antenna has a higher gain than the stated one. In comparison with the previously reported antennas, the suggested antenna provides compact and high gain for X-band applications that can be constructed in real time.

4 Conclusion

A patch array antenna with a gain of 14 dBi and a return loss of −41 dB is built for marine radar applications, and its radiation properties are investigated. This arrangement is made up of sixteen similar square elements that are grouped along the arms raised of a one-dimensional arrangement, each of which contains four elements. The cross-polarization level is greatly reduced due to the uniform structure with respect to the x-axis. When compared to a traditional patch antenna, each patch has a slit cut into it in this innovative design, and the number of patches is greatly reduced.

References

1. Mikulasek T, Georgiadis A, Collado A, Lacik J (2013) 2 microstrip patch antenna array fed by substrate integrated waveguide for radar applications. IEEE Antennas Wirel Propag Lett 12:1287–1290
2. Jin H, Chin K, Che W, Chang C, Li H, Xue Q (2015) A broadband patch antenna array with planar differential L-shaped feeding structures. IEEE Antennas Wirel Propag Lett 14:127–130
3. Ikram M, Nguyen-Trong N, Abbosh A (2020) A simple single-layered continuous frequency and polarization-reconfigurable patch antenna array. IEEE Transactions on Antennas and Propagation 68(6):4991–4996

4. Weily AR, Nikolic N (2013) Circularly polarized stacked patch antenna with perpendicular feed substrate. IEEE Trans Antennas Propag 61(10):5274–5278
5. Hu J, Hao Z-C, Hong W (2017) Design of a wideband quad-polarization reconfigurable patch antenna array using a stacked structure. IEEE Trans Antennas Propag 65(6):3014–3023
6. Jin H, Chin K, Che W, Chang C, Li H, Xue Q (2014) Differential-fed patch antenna arrays with low cross polarization and wide bandwidths. IEEE Antennas Wirel Propag Lett 13:1069–1072
7. Yoo S, Milyakh Y, Kim H, Hong C, Choo H (2020) Patch array antenna using a dual coupled feeding structure for 79 GHz automotive radar applications. IEEE Antennas Wirel Propag Lett 19(4):676–679. https://doi.org/10.1109/LAWP.2020.2976545
8. Anguera J, Puente C, Borja C, Soler J (2012) Dual-frequency broadband—stacked microstrip antenna using a reactive loading and a fractal shaped radiating edge. IEEE Antennas Wireless Propagation Letters 6:309–312
9. Balanis A (1997) Antenna theory, Chap. 14, 2nd edn. Wiley, New York
10. Yu C-C, Yang J-H, Chen C-C, Hsieh W-C (2011) A compact printed multi-band antenna for laptop applications. In: Progress in electromagnetics research symposium proceedings, Suzhou, China
11. Gai S, Jiao YC, Yang YB, Li CY, Gong JG (2010) Design of a novel microstrip-fed dual band slot antenna for WLAN application. Progress in Electromagnetic Research Letters
12. Hsu CH, Huang CL, Tsen CF (2009) Microstrip-fed monopole antenna with a short parasitic element for wideband application. Progress in Electromagnetics Research Letters 7:115–125
13. Rathore A, Nilavalan R, Abuparboush HF, Peter T (2010) Compact dual band 2.4/5.2GHz monopole antenna for WLAN applications. IEEE Propagations
14. Wasim M, Khera S (2021) Antenna array system with enhanced gain using cross dipole for the LTE. In: 2021 9th international conference on reliability, Infocom technologies and optimization (trends and future directions) (ICRITO). IEEE, pp 1–4
15. Malaisamy K, Santhi M, Robinson S (2021) Design and analysis of 4 × 4 MIMO antenna with DGS for WLAN applications. Int J Microw Wirel Technol 13(9):979–985
16. Malaisamy K, Santhi M, Robinson S et al (2021) Design and development of dipole array antenna for Wi-Fi applications. Wireless Pers Commun 121(1)
17. Venba B, Malaisamy K (2017) Modified cross dipole antenna for Ku band satellite application. International Journal of Engineering Research & Technology (IJERT) ICONNECT 5(13)
18. Malaisamy K, Santhi M, Robinson S, Mohd W, Murugapandiyan P (2020) Design and development of cross dipole antenna for satellite applications. Frequenz 74(7–8):229–237
19. Malaisamy K, Sakrika S, Sangavi M, Sangeetha B, Swetha S (2018) Wearable antenna for radar applications. International Journal of Engineering Research & Technology (IJERT) ICONNECT 6(7)

2-Odd Labelling of Graphs and Its Applications in Satellite Communication

P. Abirami, N. Srinivasan, and A. Parthiban

Abstract Chromatic graph theory provides "an efficient model for solving satellite scheduling problems. Modems are represented by nodes, and service requests between modems are represented by lines connecting the nodes. The underlying model is a multigraph, a graph with multiple edges between the same pair of nodes. A schedule corresponds to a line-colouring of the multigraph. Each colour represents a time slot, and lines sharing a node must be coloured differently so that a modem is not required to simultaneously participate in two different requests. A schedule exists if the number of colours used does not exceed the number of time slots in the frame and if the number of lines coloured the same colour never exceeds the number of carrier frequencies allowed". "2-odd labelling of a graph $G(V, E)$ with node set V is an injection 'f' from V to Z (the set of all integers) such that the absolute difference between the labels of the adjacent nodes, u and v, is either an odd number or exactly 2. If G admits 2-odd labelling, then it is called a 2-odd graph". In this paper, 2-odd labelling of some well-known graphs are derived, besides highlighting some interesting applications of graph labelling in satellite communication.

Keywords Graph labelling · 2-odd labelling · 2-odd graph · Satellite communication

1 Introduction

Graphs "considered here are all simple, finite, connected, and undirected". By $G(V, E)$, we signify the graph G having node set V and line set E. For graph theory concepts, refer to [1]. According to Laison et al. [2], "G is 2-odd if there exists an injective labelling $h : V(G) \to Z$, such that for any (uv) $\in E(G)$, $|h(u) - h(v)|$

P. Abirami · N. Srinivasan
Department of Mathematics, St. Peter's Institute of Higher Education and Research, Avadi 600054, Tamil Nadu, India

A. Parthiban (✉)
Department of Mathematics, School of Advanced Sciences, Vellore Institute of Technology, Vellore, India
e-mail: parthiban.a@vit.ac.in

D. Singh et al. (eds.), *Computer Aided Constellation Management and Communication Satellites*, Lecture Notes in Electrical Engineering 987,
https://doi.org/10.1007/978-981-19-8555-3_10

is 2 or odd integer”. They also stated that “$h(uv) = |h(u) - h(v)|$ and termed h as 2-odd labelling of G”. Furthermore, “$h(uv)$ may still be either 2 or odd if uv is not in E(G)”. For more results on 2-odd graphs, see [2–4]. A complete characterization of 2-odd labelling is an open problem of research. This paper deals with the problem partially and derives 2-odd labelling of some graphs.

2 Results and Discussion

2-odd labelling of certain graphs is established in this section.

Lemma 1 [2] *Let H be a subgraph of G. If G is not 2-odd, then H is also so.*

Definition 1 [5] "P_{Z_n}, a graph formed by subdividing each of spokes of wheel graph W_n is said to be pizza graph with $2n + 1$ nodes”.

Theorem 1 *P_{Z_n} allows 2-odd labelling for every $n \geq 3$.*

Proof “Let P_{Z_n} be a pizza graph on $n \geq 3$ nodes that consists of $2n + 1$ nodes. We name the nodes on the rim by $u_1, u_2, \ldots, u_n$, inner nodes by $v_1, v_2, \ldots, v_n$ and a central node by v_0. Now define a one-to-one function $f : V(P_{Z_n}) \rightarrow Z$ in the following way: arbitrarily, let $f(v_0) = 2n$, $f(u_i) = 2i, \forall 1 \leq i \leq n - 1$ and $f(u_n) = f(u_{n-1}) + 1$. Then $f(v_i) = 2i - 1, \forall 1 \leq i \leq n - 1$, and $f(v_n) = f(u_n) + 2$. It can be easily shown that f is the needed 2-odd labelling of P_{Z_n} as $|f(v_0) - f(v_i)|$ are all odd numbers $\forall 1 \leq i \leq n - 1$, $|f(v_i) - f(u_i)| = 1$, $|f(v_n) - f(u_n)| = 2$, $|f(u_i) - f(u_{i+1})| = 2, \forall 1 \leq i \leq n - 2$, $|f(u_n) - f(u_1)|$ and $|f(u_{n-1}) - f(u_{n-2})|$ are odd integers”. Hence, the proof. (For example, see Fig. 1).

Definition 2 “S_n, a sunlet graph, constructed by inserting $n-$ pendant lines to C_n, be on $2n$ nodes”.

Definition 3 “$L(G)$, the line graph of G, whose nodes are lines of G, and if $u, v \in E(G)$, then $uv \in E(L(G))$, if u and v share a node in G”.

Theorem 2 *$L(S_n)$ allows 2-odd labelling for $n \geq 3$.*

Proof Set $S_n, n \geq 3$ as sunlet graph on $2n$ nodes and $L(S_n)$ be line graph of S_n with $V(L(S_n)) = V_1 \cup V_2$, where $V_1 = \{u_1, u_2, \ldots, u_n\}$ represent nodes of the cycle and $V_2 = \{v_1, v_2, \ldots, v_n\}$ are the outer nodes. Describe an injective function $f : V(L(S_n)) \rightarrow Z$, in a following way. Let $f(u_1) = 0$, $f(u_i) = f(u_{i-1}) + 2; 2 \leq i \leq n - 1$, and $f(u_n) = f(u_{n-1}) + 3$. Similarly, let $f(v_1) = f(v_{i-1}) + 2; 2 \leq i \leq n - 1$ and $f(v_n) = f(v_{n-1}) + 4$. It can be easily seen that f is the needed 2-odd labelling of $L(S_n)$. (For example, see Fig. 2).

Definition 4 “BF_n, the generalized butterfly graph, formed by attaching nodes to all wings with the property that total of introducing nodes to all the wings are equal”.

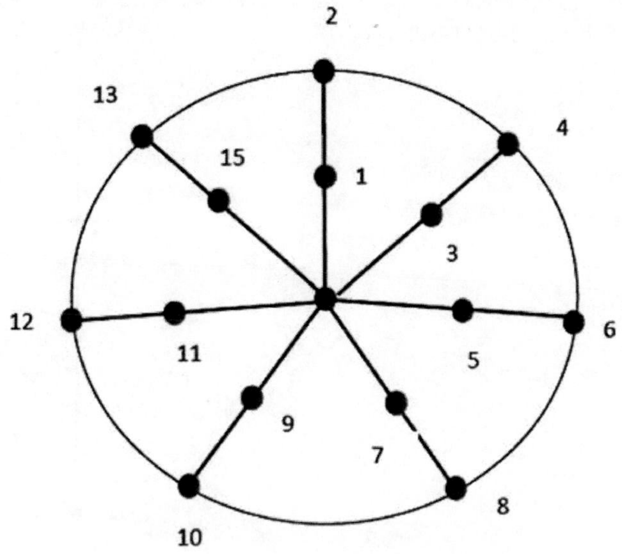

Fig. 1 2-odd labelling of P_{Z_7}

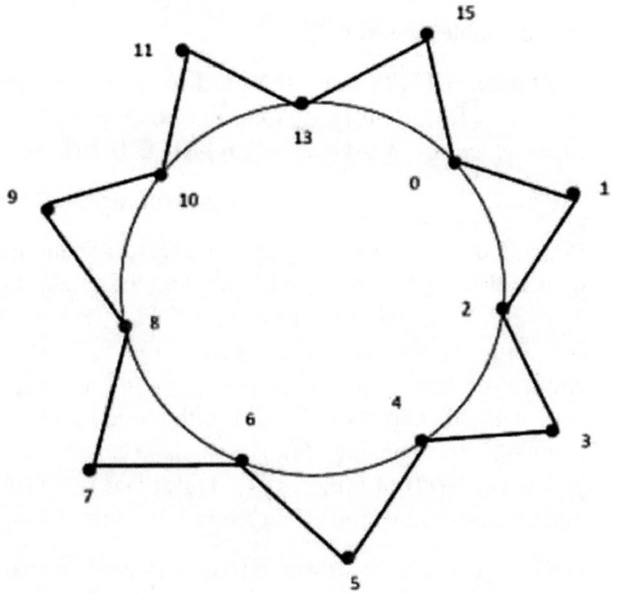

Fig. 2 2-odd labelling of $L(S_7)$

Theorem 3 *BF_n allows 2-odd labelling* $\forall\, n \geq 3$.

Proof Consider BF_n as defined above. Clearly, $|V(BF_n)| = 2n+1$ and $|E(BF_n)| = 4n - 2$. Let $V(BF_n) = V_1 \cup V_2$, where $V_1 = \{u_1, u_2, \ldots, u_n\}$ represent the right wing nodes and represent left wing nodes. Let v_0 be the apex node. Now explore an injective map $f : V(BF_n) \to Z$ in a following way. With no loss of generality, set $f(v_0) = 0$, $f(v_1) = 1$, $f(v_i) = f(v_{i-1}) + 2; 2 \leq i \leq n$, $f(u_i) = -1$ and

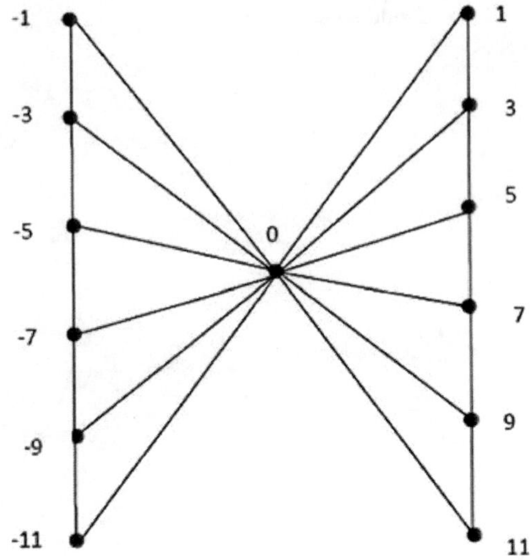

Fig. 3 2-odd labelling of BF

$f(u_i) = f(u_{i-1}) - 2; 2 \leq i \leq n$. Hence, f is the desired 2-odd labelling of BF_n. (For example, see Fig. 3).

Definition 5 [5] "Let $n \in N$ with $n \geq 3$. An Origami graph O_n is a graph with $V(O_n) = \{u_i, v_i, w_i; i \in N\}$ and $E(O_n) = \{u_i w_i, u_i v_i, v_i w_i; i \in \{1, 2, \ldots, n\}\} \cup \{u_i u_{i+1}, w_i u_{i+1}; i \in \{1, 2, \ldots, n-1\} \cup \{u_n u_1, w_n u_1\}$".

Theorem 4 *O_n, the Origami graph, allows 2-odd labelling for every $n \geq 3$.*

Proof Consider Origami graph O_n. Describe an injective function $f : V(O_n) \rightarrow Z$ in a following method. With no loss of generality, let: "$f(u_1) = 0, f(v_1) = 1, f(w_1) = 2, f(u_i) = f(u_{i-1}) + 3; 2 \leq i \leq n-1, f(v_i) = f(v_{i-1}) + 3; 2 \leq i \leq n-1$, and $f(w_i) = f(w_{i-1}) + 3; 2 \leq i \leq n-1$". Then one has to choose the labels for the nodes u_n, v_n, w_n in such a way that the absolute difference with other adjacent labels is either odd number or exactly 2, respectively, which is possible by choosing, $f(u_n) = -2$, $f(v_n) = -1$, and $f(w_n) = -3$. Undoubtedly, "$|f(u_i) - f(v_i)|, |f(u_i) - f(u_{i+1})|, |f(v_i) - f(w_i)|, |f(u_i) - f(w_i)|$" are either odd or exactly 2. Hence, f is 2-odd labelling of Origami graph.

Definition 6 [5] "A flower $F(C_m, C_n)$ is produced by taking a copy of C_m and m copies of C_n and attaching i th copy of C_n at i th line of C_m".

Theorem 5 *$F(C_m, C_n)$ allows 2-odd labelling $\forall m, n \geq 3$.*

Proof Take $F(C_m, C_n)$, a flower graph, as defined above. Explore injective function $f : V(F(C_m, C_n)) \rightarrow Z$ in the following way: with no loss of generality, let $f(u_1)$ or $f(v_1^1) = 0$, $f(v_1^1) = 2$, and $f(v_1^1) = f(v_{i-1}^1) + 2; 2 \leq i \leq n-1$. Then $f(v_n^1)$ or $f(u_2) = f(v_{n-1}^1) + 1$. Similarly, let $f(v_{n-1}^1) + 2$ and $f(v_1^2) = f(v_{i-1}^2)$ $+2, \forall 3 \leq i \leq n-1$. Then $f(u_3) = f(v_{n-1}^2) + 1$. Continue the same process, in such a way that $|f(v_i^i) - f(v_{i+1}^i)|, |f(u_i) - f(u_{i+1})|$ are either odd or exactly 2. One can see that $f(v_{n-1}^n) = f(v_{n-2}^n) + 1$ in order to get the odd distance.

Definition 7 [5] "Take $F_n = P_n + \{x\}$, where $P_n = \{v_1, v_2, \ldots, v_n\}$ is a path and for every $i \in [1, n]$, x is adjacent to v_i. A flower (C_3, F_n), obtained by considering 1 copy of C_3 and three copies of F_n and by adding the $(v_1, x) \in E(F_n)$ in each line of C_3".

Theorem 6 *$F(C_3, F_n)$ allows 2-odd labelling for $n \geq 1$.*

Proof Take $F(C_3, F_n)$ as defined above. Define an injective function $f : V(F(C_3, F_n)) \rightarrow Z$ in the following way: with no loss of generality, let $f(u_1) = 0$, $f\left(u_i^{1,2}\right) = 3$, $f\left(u_i^{1,2}\right) = f\left(u_{i-1}^{1,2}\right) + 2; 2 \leq i \leq n$, $f(u_2) = 2$, $f(u_3) = f(u_n^{1,2})+$, $f\left(u_1^{2,3}\right) = f\left(u_{i-1}^{2,3}\right) + 2; 2 \leq i \leq n$, and $f(u^{3,1}) = f(u_3) + 2l$, where $2l$ is the sufficiently large even so that $f(u^{3,1}) = f\left(u_{i-1}^{3,1}\right) + 2; 2 \leq i \leq n$. Thus, f is 2-odd labelling of $F(C_3, F_n)$.

Definition 9 [6] "The graph $Pl_n = (V, E)$, where $V = \{1, 2, \ldots, n\}$ and $E = E(K_n)\backslash\{(k, l)3 \leq k \leq n-2$ and $k+2 \leq l \leq n\}$ is a planar graph having maximum number of lines, with n nodes. The planar graph Pl_n having maximum number of lines with n nodes is obtained by removal of $[(n-4)(n-3)]/2$ lines from K_n. The number of lines in $Pl_n : n \geq 5$ is $3(n-2)$". One such example Pl_{10} class graph is shown in Fig. 4.

Theorem 8 *The planar graph Pl_n allows 2-odd labelling.*

Proof "Let Pl_n be the given planar graph with $V(Pl_n) = V_1 U V_2$, where $V_1 = \{u_1, u_2\}$ and $V_2 = \{v_i; 1 \leq i \leq n\}$. Now define an injective function $f : V(Pl_n) \rightarrow$

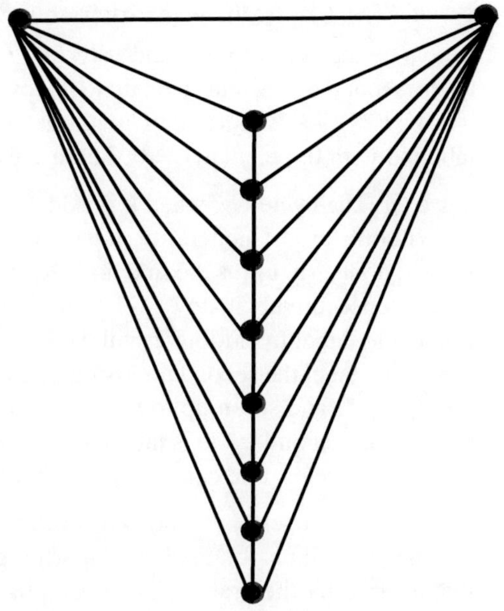

Fig. 4 Planar graph Pl_{10}

Fig. 5 Friendship graph

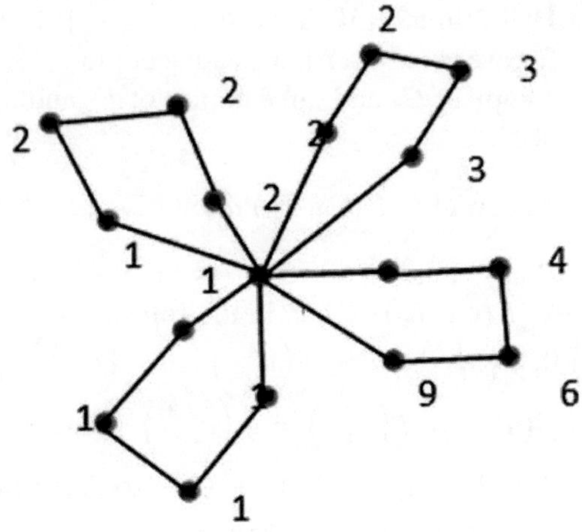

Z as follows: without loss of generality, let $f(u_1) = 1$ and $f(u_2) = -1$. Then $f(v_i) = 2i; 1 \le i \le n$. One can see that $|f(u_1) - f(u_2)| = 2$ and $|f(u_1) - f(v_i)|$ are odd numbers and $|f(u_2) - f(v_i)|$ are also odd numbers for all $1 \le i \le n$". Hence, f is the required 2-odd labelling of Pl_n.

Definition 11 [7] "The friendship graph T_n is a set of n triangles having a common central node".

Definition 12 [7] "The friendship graph $F_{r_n}^{(k)}$ is a planar undirected graph with $(k-1)n+1$ nodes and kn lines connected by joining n copies of the cycle C_k with a common node". (See Fig. 5).

Theorem 10 *The friendship graph $F_{r_n}^{(k)}$ allows 2-odd labelling for all $k \ge 3$.*

Proof "Let $F_{r_n}^{(k)}$ be the given friendship graph on $(k-1)n+1$ nodes and kn lines. Let v_0 be the central node and $v_i^j : 1 \le i \le k, 1 \le j \le n$ be the nodes on the first cycle, second cycle, and so on up to n^{th} cycle". Now we define a one-to-one labelling $f : V(F_{r_n}^{(k)}) \to Z$ as follows: without loss of generality, let the central node v_0 be labelled with 0, i.e. $f(v_0) = 0$. There arise two cases.

Case (i) The cycle C_k, when k is odd.

Without loss of generality, label the nodes on the first cycle, starting with the second node, say v_1 (as the first node v_0 is already labelled) by giving 1, for v_2 assign 2, and consecutively up to $(n-1)$th node assign $n-1$. Finally, for the nth node in the first cycle, label by adding 2 with $(n-1)$th label. That is $f\left(v^1{}_n\right) = f\left(v^1{}_{n-1}\right) + 2$. Similarly, label the second node of the second cycle with $f\left(v^1{}_n\right)+2$, third node with $f\left(v^2{}_2\right) + 1$, and so on up to the $(n-1)$th node of the second cycle. Finally, the nth node of the second cycle is labelled with $f\left(v^2{}_{n-1}\right) + 2$. Proceeding the same for the remaining cycles, one can easily see that "f is the required 2-odd labelling of $F_{r_n}^{(k)}$".

Case (ii) The cycle C_k, when k is even.

The graph thus obtained is a bipartite graph, and the result follows from the fact that every bipartite graph is 2-odd graph.

3 Applications

Graph colouring (node colouring) is "an assignment of colours to the nodes of G such that the adjacent nodes receive distinct colours. Similarly, line colouring is an assignment of colours to the lines of G such that the adjacent lines receive distinct colours. The general modem-to-modem satellite communications scheduling problem is purely a problem of minimizing the number of colours used to line colour a multigraph. As long as the area constraint has been met, there is no need to worry about restricting the number of occurrences of a given colour to the maximum number of available carrier frequencies. The reason, apparently first observed independently by [8] and [9], is that any line colouring of a multigraph using n colours can be transformed into an n-colouring which is as balanced as possible. If, for instance, colour A occurs at least two more times than colour B, consider the subgraph determined by the line coloured with A and B. This subgraph must be a disjoint collection of chains and even-length cycles of lines alternating in the colours A and B, and so there must be an odd length chain in which the colour A appears one more time than B does. Reversing the colours on this chain brings the colouring more into balance and iterating this process until no colour appears two or more times than other yields a balanced colouring". For a detailed study, refer to [10].

4 Conclusion

2-odd labelling of some particular simple graphs like flower graphs, sunlet, Origami, Pizza graph, jewel graph, planar graph, Mongolian tent graph, friendship graphs are derived in this paper. A complete characterization of 2-odd graphs is still an open question and which may be carried out for future research. We also believe that 2-odd labelling may find its exclusive applications in designing satellites and their communications.

References

1. Diestel R (2005) Graph theory. Springer, Berlin
2. Laison JD, Starr C, Walker A (2013) Finite prime distance graphs and 2-odd graphs. Discret Math 313:2281–2291
3. Abirami P, Parthiban A, Srinivasan N (2020) On 2—odd labeling of graphs. European Journal of Molecular and Clinical Medicine 07:3914–3918
4. Abirami P, Parthiban A, Srinivasan N (2020) Some results on 2-odd labeling of graphs. International Journal of Recent Technology and Engineering (IJRTE) 8(5):5644–5646
5. Nabila, S., Salman, A.N.M: The rainbow connection number of origami graphs and pizza graphs. Procedia Computer Science, 74, 162 – 167, (2015).
6. Baskar Babujee J (2003) Plannar graphs with maximum-edges antimagic property. Math Ed 37(4):194–198

7. Meena S, Vaithilingam K (2012) Prime labelling of friendship graphs. International Journal of Engineering Research and Technology (IJERT) 1(10)
8. Diarmid MC (1972) The solution of a time-tabling problem. J Inst Malhs 9:23–34
9. De Werra D (1971) Equitable colorations of graphs. Rev Fran In! Rech Oper 5:3–8
10. Bean DR, Engel GM (1992) Satellite communications and multigraph edge-coloring. Australasian Journal of Combinatorics 5:43–52

Comparative Analysis of Different Dielectric Substrate for the Design of Millimeter Wave Microstrip Patch Antenna

Reena Aggarwal, Ajoy Roy, and Gurpreet Kumar

Abstract This work gives the comparison of different substrate materials required for designing a microstrip patch antenna, i.e., FR-4, RO4003C, Arlon DiClad and RT Duroid that operate in the frequency band of 22–30 GHz. The purpose of this work is to compare and propose a particular substrate that can be used in a patch antenna at Ka frequency band. The height of the substrate material is kept fixed at 1.6 mm for all substrates. Simulation-based comparative analysis is done on different dielectric substrate materials to evaluate the antenna metrics such as antenna input impedance, VSWR, S-parameters, gain, and radiation efficiency.

Keywords Microstrip patch antenna · Dielectric substrate · FR-4 · RT Duroid · RO4003C · Antenna gain

1 Introduction

Because of the low profile and simple in fabrication, microstrip patch antenna is widely used antenna structure. It is most acceptable antenna structure in a large portion of the field because of their low volume, lightweight, and planar design which can made conformal. They are utilized as a part of numerous applications, for example, wireless and in satellite communication. But on the other hand it takes a shot at the thin data transmission. To plan microstrip patch antenna needs prerequisites shape, dimensions, feeding, and operating frequency. But the most critical prerequisite is determination of substrate in view of the cost, effectiveness, and size.

Substrates utilized for microstrip patch antenna lie between $2.2 \leq \varepsilon \leq 12$. Lower the permittivity of dielectric material higher the gain of antenna. There are wide groupings of material for substrates have been found for MSPA with thermal, mechanical and electrical properties, it is imperative to know the effect of varying

R. Aggarwal · A. Roy (✉) · G. Kumar
Lovely Professional University, Phagwara 144402, India
e-mail: ajoy.22652@lpu.co.in

R. Aggarwal
e-mail: reena.16120@lpu.co.in

D. Singh et al. (eds.), *Computer Aided Constellation Management and Communication Satellites*, Lecture Notes in Electrical Engineering 987,
https://doi.org/10.1007/978-981-19-8555-3_11

dielectric substrate material and substrate thickness. MSPA in comparison with conventional antennas offer several advantages. Research in the design of microstrip patch antenna has been done by several researchers by varying the dielectric substrate material and its thickness. In [1–11], antenna was designed using RT Duroid substrate to achieve desired antenna parameters. The use of Rogers 5880 substrate provides high gain and bandwidth with lower dielectric constant. The researchers in [12] and [13] designed antenna using FR-4 substrate. The designed antenna in [12] was not considered wideband and provided a total efficiency >60%. In [13], the mutual coupling between adjacent antennas observed was less than −17.5 dB at resonance frequency of 34.6 GHz. Researchers in [14–16] designed antenna on a Rogers 4003C substrate. Selection of the correct dielectric substrate becomes a challenge for the researchers to decide, which of the common available substrates will provide the better antenna performance, thereby four different substrate materials are compared to know their dielectric properties that affect the performance of antenna.

In this study, four different substrate materials are reviewed with their respective properties. Different substrates such as FR-4, RO 4003C, Arlon DiClad, and RT Duroid are compared by keeping substrate height constant at 1.6 mm.

2 Dielectric Substrate

A dielectric substrate acts as an insulator, lesser the value of dielectric constant higher the bandwidth because better fringing is achieved with lower relative permittivity. Different substrate such as FR-4, RO4003C, Arlon DiClad and RT Duroid are used in designing a microstrip patch antenna for 22–30 GHz frequency range. The thickness of the dielectric substrate is kept fixed at 1.6 mm for all substrates. A thick dielectric substrate with a low dielectric constant is preferable, thereby to get high radiation efficiency and wide bandwidth. FR-4 is a most commonly used composite material with considerable mechanical strength, having almost zero water absorption. It is a less expensive material. RO4003 substrates are manufactured from standard epoxy/glass (FR-4) forms. These low cost substrates are considered as a low loss material. Rogers DiClad 870 and 880 laminates are woven fiberglass reinforced, PTFE-based composites that offer lower dielectric constant for use as printed circuit board substrates. It offers improved dissipation factor without sacrificing mechanical properties. RT Duroid is glass microfiber composite manufactured by Roger Corporation. They exhibit uniform electrical properties and excellent chemical resistance. Table 1 gives the properties of different types of substrate materials.

Table 1 Properties of different dielectric substrate

Parameters	FR-4	RO4003C	Arlon DiClad	RT Duroid
Dielectric constant (ε_r)	4.36	3.4	2.33	2.2
Loss tangent	0.013	0.002	0.0005	0.0004
Water absorption	<0.25%	0.06%	0.02%	0.02%
Tensile strength	<310 MPa	141 MPa	<105 MPa	450 MPa
Volume resistivity	8 × 107 MΩcm	1700 × 107 MΩcm	1.5 × 109 MΩcm	2 × 107 MΩcm
Surface resistivity	2 × 1015 MΩ	4.2 × 109 MΩ	3.4 × 107 MΩ	3 × 107 MΩ
Breakdown voltage	55 kV	–	–	>60 kV
Peel strength	9 N/mm	1.05 N/mm	2.4 N/mm	5.5 N/mm
Density	1850 kg/m^3	1790 kg/m^3	1260 kg/m^3	2200 kg/m^3

3 Design of Microstrip Patch Antenna

3.1 Microstrip Patch Antenna

A rectangular shaped MSPA, as shown in Fig. 1, with dimensions of L × W is etched over a substrate of dielectric constant εr and height h. To design the single patch antenna, there are three critical parameters:

- Operating frequency, fr = 26 Ghz
- Substrate height, $h = 1.6$ mm
- Dielectric constant of the substrate, 2.2 ≤ ε ≤ 12

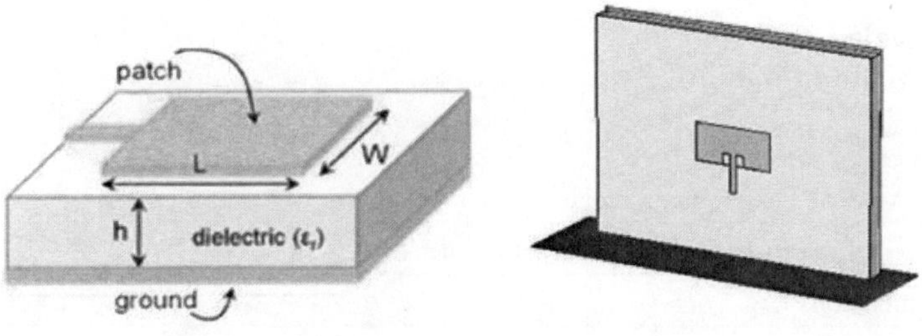

(a) (b)

Fig. 1 Design of microstrip patch antenna **a** top view **b** side view

3.2 Equations

Dimensions of microstrip patch antenna, viz. its width and length is calculated using the formulas mentioned below in Eqs. (1)–(3).

$$W = \frac{c}{2f\sqrt{\left(\frac{\varepsilon_r+1}{2}\right)}} \tag{1}$$

$$L = \frac{c}{2f\sqrt{\varepsilon_{r_{\text{eff}}}}} - 0.824h\frac{\left(\left(\varepsilon_{r_{\text{eff}}} + 0.3\right)\left(\frac{w}{h} + 0.264\right)\right)}{\left(\left(\varepsilon_{r_{\text{eff}}} - 0.258\right)\left(\frac{w}{h} + 0.8\right)\right)} \tag{2}$$

$$\varepsilon_{r_{\text{eff}}} = \frac{\varepsilon_r + 1}{2} + \frac{\varepsilon_r - 1}{2}\left(1 + 12\frac{h}{w}\right)^{-1/2} \tag{3}$$

This effective dielectric constant in Eq. (3) accounts for the fringing effect, basically due to the differences in the phase velocities of air and the substrate. Due to which the fringing fields around the boundary surface of the patch are not confined in the dielectric substrate but they also spread in the air. Equation (4) gives the expression for calculating the feed length of an antenna, which in turn is calculated from the guided wavelength mentioned in Eq. (5). Table 2 shows the dimensions of a patch antenna at 26 GHz frequency.

$$L_f = \frac{\lambda_g}{4} \tag{4}$$

$$\lambda_g = \frac{\lambda}{\sqrt{\varepsilon_{r_{\text{eff}}}}} \tag{5}$$

where

λ wavelength
c velocity of light (3 × 1011 mm)

Table 2 Antenna design parameters

Parameters of microstrip patch antenna in mm		
Patch dimensions		
$L = 1.965$	W = 3.509	Hp = 0.035
Ground dimensions		
Gl = 11.731		Gw = 13.144
Substrate dimensions		
Sl = 11.367		Sw = 13.144

where Hp = Patch Thickness, Gl = Ground Length, Gw = Ground Width, Sl = Substrate Length, and Sw = Substrate Width

f Resonant frequency
ε_r Dielectric constant of a substrate
$\varepsilon_{r_{\text{eff}}}$ effective dielectric constant (accounts for fringing effect, $\varepsilon_{r_{\text{eff}}} < \varepsilon_r$)
w patch width
h substrate height
L patch length

4 Performance Metrics of Patch Antenna

4.1 Return Loss

It is the measure of the extent an antenna is matched to the 50-Ω transmission line. Under the conditions of impedance mismatching, it calculates the amount of the incident power being reflected as given in Eq. (6). This section provides the simulated results of return loss for different substrate materials.

$$RL(\text{dB}) = 10\log\left(\frac{P_{\text{incident}}}{P_{\text{reflected}}}\right) \tag{6}$$

Return loss can be calculated from VSWR by the relation specified in Eq. (7).

$$RL = -20\log\frac{\text{VSWR} - 1}{\text{VSWR} + 1} \tag{7}$$

4.2 Voltage Standing Wave Ratio (VSWR)

Voltage standing wave signifies the amount of power reflection from the antenna. In the case of patch antenna, VSWR value should not be more than 2 and less than 1 along the bandwidth of efficiency. Ideally, it should be 1. Equation (8) relates the reflection coefficient with VSWR. This section shows the simulation results of different substrate materials in terms of VSWR.

$$\text{VSWR} = \frac{1 + |\Gamma|}{1 - |\Gamma|} \tag{8}$$

4.3 Antenna Gain

The microstrip patch antenna was designed for different substrate materials by keeping substrate height fixed. From Figs. 2, 3, 4 and 5, it has been observed clearly that the gain of patch antenna is high when designed with RT Duroid substrate material with relative permittivity of 2.2. As it can be seen from the graph, the antenna is designed for broadband applications varying from 22 to 30 GHz. FR-4 substrate provides the least gain at a resonant frequency of 26 GHz. Hence, RT Duroid substrate is preferred for high gain applications compared to FR-4, Rogers, and Arlon.

FR-4

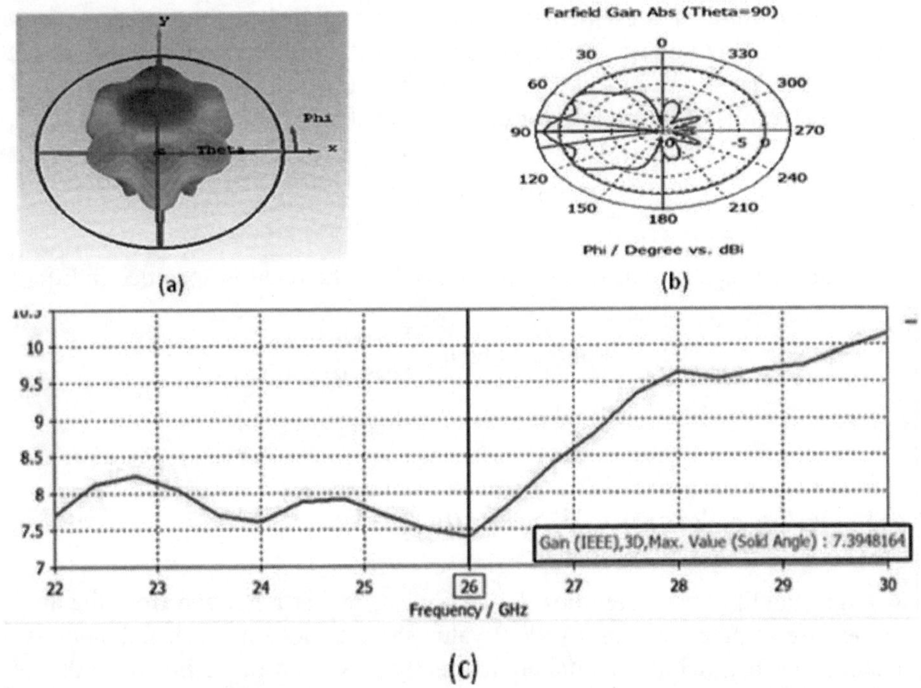

Fig. 2 Antenna gain for FR-4 substrate **a** fairfield 3D antenna gain **b** antenna gain in polar coordinates and **c** gain versus frequency graph

RO 4003C.

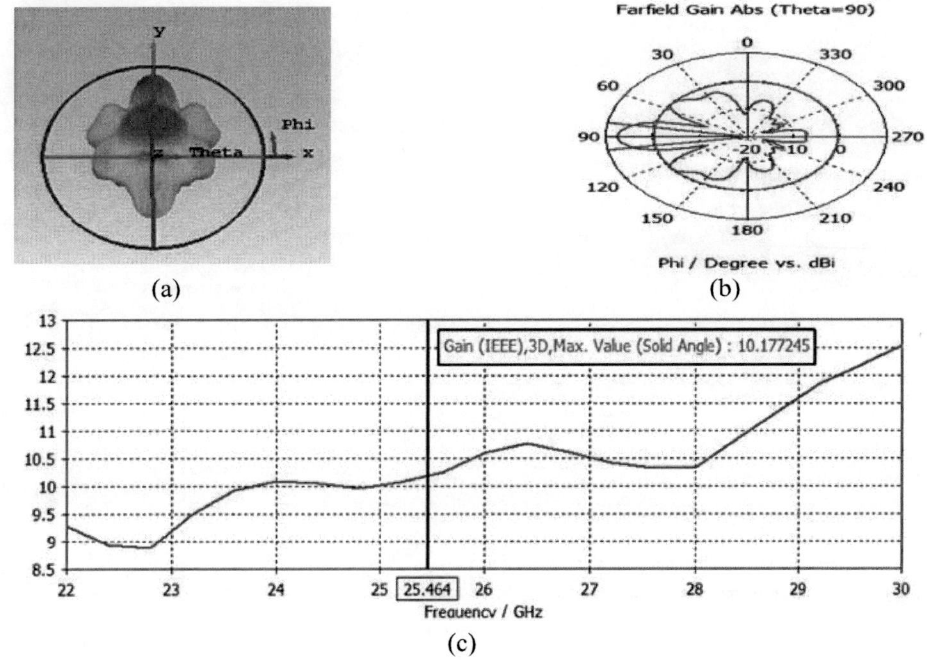

Fig. 3 Antenna gain for RO4003C substrate **a** fairfield 3D antenna gain **b** antenna gain in polar coordinates and **c** gain versus frequency graph

5 Simulation-Based Comparison of Different Dielectric Substrates

5.1 S11 or Return Loss

From Fig. 6, it has been observed that FR-4 substrate material provides maximum return loss of about 35.53 dB. Also, use of a RO4003C as a substrate gives the poor return loss.

5.2 VSWR

In the case of patch antenna, VSWR should not be more than 2 or less than 1 along the bandwidth of efficiency. This corresponds to a return loss (RL) of approximately −10 dB. It can be seen from Fig. 7 that the VSWR of all the substrates are within the accepted limits, with FR-4 substrate providing better value compared to other substrates materials.

Arlon DiClad.

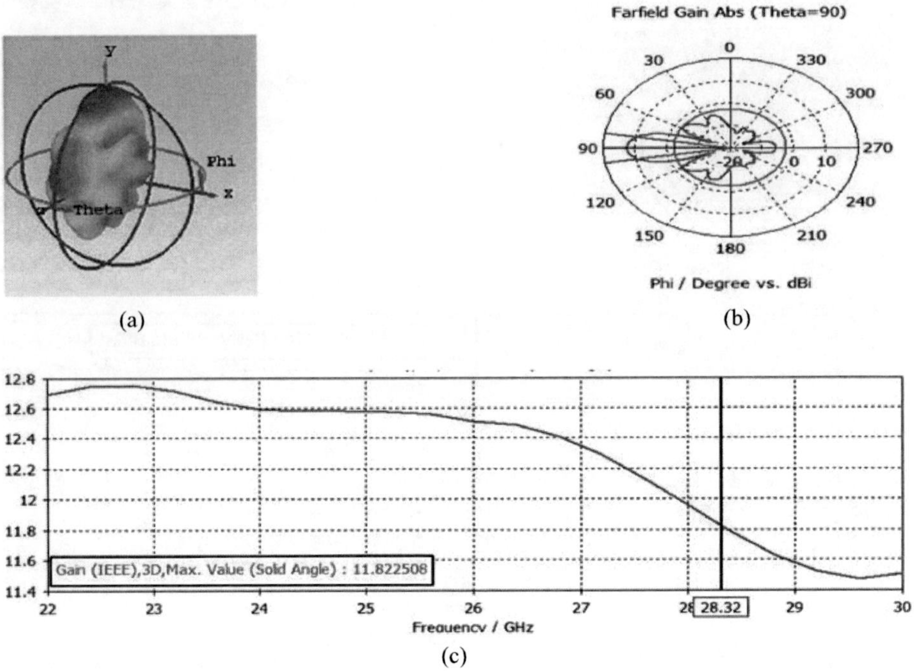

Fig. 4 Antenna gain for Arlon substrate **a** fairfield 3D antenna gain **b** antenna gain in polar coordinates and **c** gain versus frequency graph

5.3 Antenna Gain

From Fig. 8, it can be seen that RT Duroid 5880 provides the best results in terms of antenna gain as compared to other substrate materials.

5.4 Antenna Radiation Efficiency

It is a measure of how suitably a radio antenna converts the power accepted at its terminals into radiated power. Figure 9 shows that RT Duroid proves best in term of radiation efficiency, as it provides radiation of 83.5% compared to all dielectric substrate materials. The radiation efficiency of Arlon substrate is comparable to RT Duroid material.

Efficiency of an antenna is given by

$$\eta = \frac{\text{gain}}{\text{directivity}} * 100\% \tag{9}$$

RT Duroid.

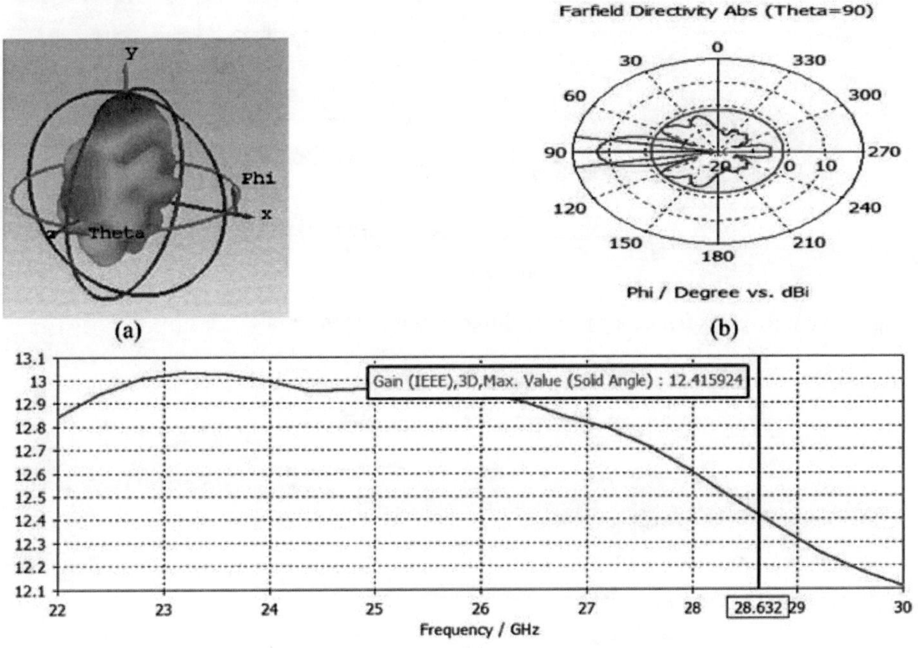

Fig. 5 Antenna gain for RT Duroid substrate **a** fairfield 3D antenna gain **b** antenna gain in polar coordinates and **c** gain versus frequency graph

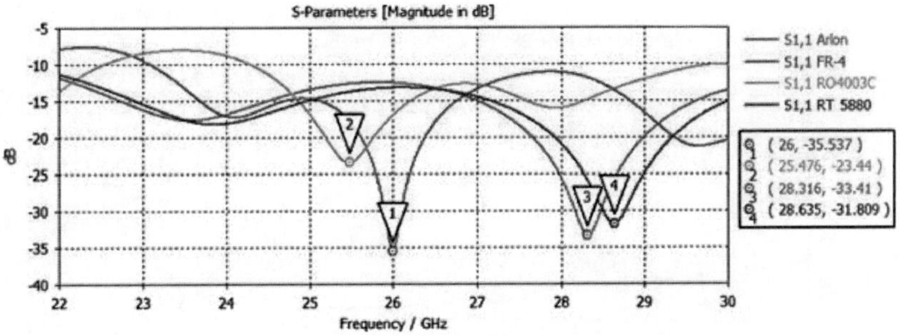

Fig. 6 Comparison among different substrate materials in terms of S11 or return loss

After calculating the efficiency of antenna using formula given in Eq. (9) for different substrates based on the values of gain and directivity mentioned in the Table 3, it has been found that the efficiency of antenna in case of RT Duroid substrate is much higher than all other substrate materials.

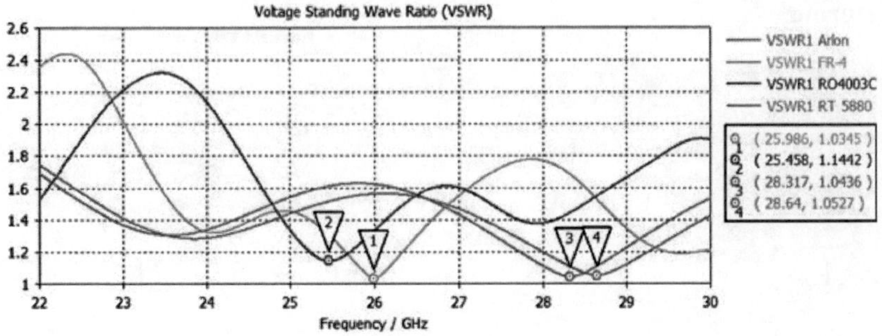

Fig. 7 VSWR versus frequency plot for different substrate materials

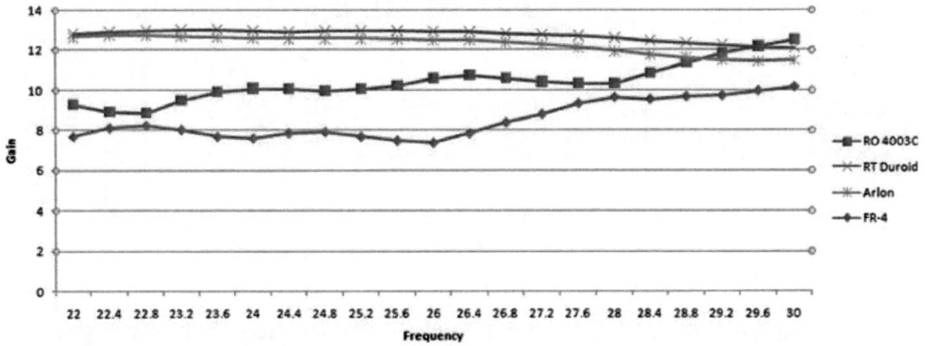

Fig. 8 Antenna gain versus frequency plot for different substrate materials

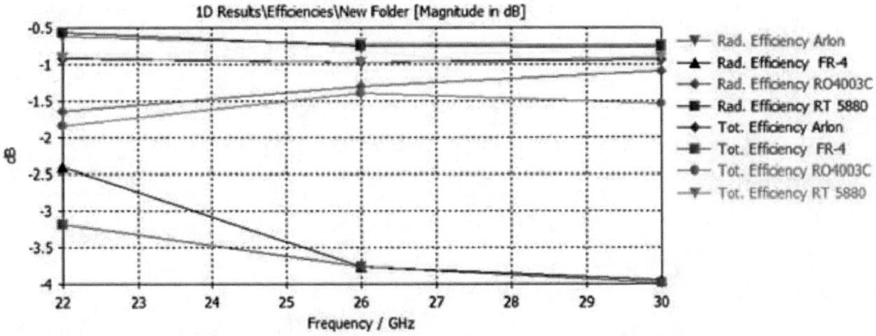

Fig. 9 Comparison among different substrate materials in terms of antenna

6 Conclusion

Antenna characteristics are affected mostly by the shape of its patch. By suitable selection of substrate thickness and patch width W_p, a broadband antenna can be designed. Four unique substrate materials FR-4, RO4003C, Arlon DiClad 870, and RT Duroid 5880 are considered for the design of the microstrip patch antenna. Each

Table 3 Simulation results of different substrate materials for the design of microstrip patch antenna

Parameters	Substrate type			
	FR-4	RO4003C	Arlon DiClad	RT Duroid
Frequency (GHz)	22	22	22	22
Resonant frequency (GHz)	26	25.46	28.32	28.63
Return loss (dB)	−35.53	−23.45	−33.42	−31.81
Gain (dBi)	7.395	10.17	11.82	12.41
Directivity (dBi)	11.15	11.88	13.23	13.70
VSWR	1.03	1.14	1.04	1.05
Antenna efficiency (dB)	−3.77	−1.27	−0.81	−0.78
Radiating efficiency (dB)	−3.77	−1.24	−0.81	−0.78

of them resonates at different frequency. RT Duroid substrate in contrast to other substrates has high antenna gain and higher antenna efficiency. The reason behind this high value of gain is because of the least permittivity of the Duroid material. RT Duroid gives most extreme radiation because of its low dielectric value. Bandwidth and radiation design relies upon loss tangent also, so RT Duroid is great dielectric substrate for microstrip patch antennas over other three substrates.

References

1. Hussain N, Jeong M-J, Abbas A, Kim N (2020) Metasurface-based single-layer wideband circularly polarized MIMO antenna for 5G millimeter-wave systems. IEEE Access 8:130293–130304
2. Malviya L, Parmar A, Solanki D, Gupta P, Malviya P (2020) Highly isolated inset-feed 28 GHz MIMO-antenna array for 5G wireless application. Procedia Computer Science 171:1286–1292
3. Nigam H, Mathur M, Arora M, Singh R (2020) Enhancing gain with multiple beams for four element phased array microstrip antenna using Rogers substrate material for 5G applications. Materials Today: Proceedings, p S2214785320345570
4. Yang B, Yu Z, Dong Y, Zhou J, Hong W (2017) Compact tapered slot antenna array for 5G millimeter-wave massive MIMO systems. IEEE Trans Antennas Propagat 65(12):6721–6727
5. Mahmoud KR, Montaser AM (2017) Optimised 4 × 4 millimetre-wave antenna array with DGS using hybrid ECFO-NM algorithm for 5G mobile networks. IET Microwaves Antennas Propag 11(11):1516–1523
6. Aliakbari H, Abdipour A, Mirzavand R, Costanzo A, Mousavi P (2016) A single feed dual-band circularly polarized millimeter-wave antenna for 5G communication. In: 2016 10th European Conference on Antennas and Propagation (EuCAP), Davos, Switzerland, pp 1–5
7. Al Abbas E, Ikram M, Mobashsher AT, Abbosh A (2019) MIMO antenna system for multi-band millimeter-wave 5G and wideband 4G mobile communications. IEEE Access 7:181916–181923
8. Marzouk HM, Ahmed MI, Shaalan AA (2019) A novel dual-band 28/38 GHz slotted microstip MIMO antenna for 5G mobile applications. In: 2019 IEEE international symposium on antennas and propagation and USNC-URSI radio science meeting, Atlanta, GA, USA, pp 607–608

9. Jilani SF, Alomainy A (2018) Millimetre-wave T-shaped MIMO antenna with defected ground structures for 5G cellular networks. IET Microwaves Antennas Propag 12(5):672–677
10. Tu DTT, Thang NG, Ngoc NT, Phuong NTB, YemVV (2017) 28/38 GHz dual-band MIMO antenna with low mutual coupling using novel round patch EBG cell for 5G applications. In: 2017 international conference on advanced technologies for communications (ATC), Quy Nhon
11. Zhang Y, Deng J-Y, Li M-J, Sun D, Guo L-X (2019) A MIMO dielectric resonator antenna with improved isolation for 5G mm-wave applications. Antennas Wirel Propag Lett 18(4):747–751
12. Li Y, Sim C-Y-D, Luo Y, Yang G (2019) High-isolation 3.5 GHz eight-antenna MIMO array using balanced open-slot antenna element for 5G smartphones. IEEE Trans Antennas Propagat 67(6):3820–3830
13. Labidi M, Choubani F (2019) A design of metamaterials MIMO antenna for millimeter wave application. In: 2019 international conference on software, telecommunications and computer networks (SoftCOM), Split, Croatia
14. Shen X, Liu Y, Zhao L, Huang G-L, Shi X, Huang Q (2019) A miniaturized microstrip antenna array at 5G millimeter-wave band. Antennas Wirel Propag Lett 18(8):1671–1675
15. Saad AAR, Mohamed HA (2019) Printed millimeter-wave MIMO-based slot antenna arrays for 5G networks. AEU—International Journal of Electronics and Communications 99:59–69
16. Hsu Y-W, Huang T-C, Lin H-S, Lin Y-C (2017) Dual-polarized quasi Yagi-Uda antennas with endfire radiation for millimeter-wave MIMO terminals. IEEE Trans Antennas Propagat 65(12):6282–6289

Analyzing the Applications of Graph Theory in Communication Networks Through the Divisor 3-equitable Labeling of Graphs

Sangeeta and A. Parthiban

Abstract The field of graph theory, specifically graph labeling is used in communication networks, particularly in satellite communication. An allocation of numbers to the nodes of a graph $G(V, E)$ under some conditions is a node labeling of$G$. In a "mobile satellite service (MSS)" system, "channel using efficiency is still the main factor" because the systems frequency reuse factor cannot reach 1 exactly. So, researchers have proposed a channel reallocation method using the concept of matching theory to handle this situation by reducing the interference level. A "divisor 3-equitable labeling is a bijective

$$\text{map}\, d : V(G) \to \{1, 2, \ldots n\} \,\text{defined by}\; \mathrm{d}(e = xy) = \begin{cases} 1, & \text{if}\, d(x)|\mathrm{d}(y) \text{ or } \mathrm{d}(y)|\mathrm{d}(x) \\ 2, & \text{if}\, \frac{\mathrm{d}(x)}{\mathrm{d}(y)} = 2 \text{ or } \frac{d(y)}{d(x)} = 2 \\ 0, & \text{otherwise} \end{cases}$$

such that $|e_d(i) - e_d(j)| \leq 1 \;\forall\; 0 \leq i, j \leq 2$, where $e_d(i)$ is the count of lines labeled with i under d. A graph that accepts divisor 3-equitable labeling is called a divisor 3-equitable graph". This article shows the "existence and non-existence of divisor 3-equitable labeling of certain graphs, besides recalling a few applications of graph labeling in communication networks".

Keywords Graph labeling · Divisor 3-equitable labeling · Degree splitting graph · Mobile satellite service

---

Sangeeta
Department of Mathematics, Lovely Professional University, Phagwara 144411, Punjab, India

A. Parthiban (✉)
Mathematics Division, School of Advanced Sciences, Vellore Institute of Technology, Vellore, India
e-mail: parthiban.a@vit.ac.in

D. Singh et al. (eds.), *Computer Aided Constellation Management and Communication Satellites*, Lecture Notes in Electrical Engineering 987,
https://doi.org/10.1007/978-981-19-8555-3_12

1 Introduction

All graphs are "finite, undirected, simple, and connected" in this present study. Let $G(V, E)$ or G denote a graph G with "node set V and line set E". Graph labeling is used in coding theory particularly in "missile guidance codes, design of good radar type codes, and convolution codes".

1.1 Preliminaries

Definition 1 [1] A "ternary node labeling f of G becomes 3-equitable labeling" if "$|v_f(i) - v_f(j)| \leq 1$ and $|e_f(i) - e_f(j)| \leq 1''$ $\forall\, 0 < i, j > 2$. A graph that admits "3-equitable labeling" is a "3-equitable graph".

Definition 2 [1] A "divisor 3-equitable labeling" is a bijective map $d : V(G) \rightarrow \{1, 2 \ldots, n\}$ defined by

$$\text{“d}(e = xy) = \begin{cases} 1, \text{ if } \mathrm{d}(x)|\mathrm{d}(y) \text{ or } \mathrm{d}(y)|\mathrm{d}(x) \\ 2, \text{ if } \frac{\mathrm{d}(x)}{\mathrm{d}(y)} = 2 \text{ or } \frac{d(y)}{d(x)} = 2 \\ 0, \text{ otherwise”} \end{cases}$$

such that "$|e_d(i) - e_d(j)| \leq 1$" $\forall\, 0 \leq i, j \leq 2$, where $e_d(i)$ is the count of lines labeled i under d. A graph that accepts "divisor 3-equitable labeling" is a "divisor 3-equitable graph".

Definition 3 [2] Let H be with $V = S_1 \cup S_2 \cup \ldots \cup S_i \cup T$, where each S_i is a set of nodes having a minimum of 2 nodes of the same degree and $T = V \backslash \cup S_i$. The degree splitting graph of H, $DSG(H)$ is constructed from H by introducing nodes $w_1, w_2, \ldots, w_t$ and connecting each node of S_i, $1 \leq i \leq t$.

This article shows the existence and non-existence of "divisor 3-equitable labeling" of the DSG of certain graphs, besides recalling a few applications of graph theory in communication networks. Let D3EL and D3EG denote "divisor 3-equitable labeling" and "divisor 3-equitable graph", respectively.

2 Results and Discussion

Theorem 1 [3] *W_n does not permit D3EL $\forall\, n \geq 7$.*

Theorem 2 [4] *K_n does not permit D3EL $\forall\, n \geq 5$.*

Theorem 3 [4] *$K_{1,n}$ does not permit D3EL $\forall\, n \geq 6$.*

2.1 D3EL of Degree Splitting Graph of Ladder Graph

Definition 4 [5] The ladder L_n is defined as: $L_n = P_n \times P_2$.

The D3EL of $DSG(L_1)$, $DSG(L_2)$, and$DSG(L_3)$ is derived easily (see Fig. 1). So, consider $DSG(L_n), \forall\ n \geq 4$.

Theorem 4 *$DSG(L_n)$ does not allow D3EL $\forall\ n \geq 4$.*

Proof Take L_4 for the sake of discussion with $|V(L_n)| = 8$ and $|E(L_n)| = 10$. Obtain $DSG(L_n)$ with $|V(DSG(L_n))| = 10$ and $|E(DSG(L_n))| = 18$. Now define a bijective map $d : V(DSG(L_n)) \to \{1, 2, \ldots, n\}$ as follows:

$$\mathrm{d}(e = xy) = \begin{cases} 1, \text{ if } \mathrm{d}(x)|\mathrm{d}(y) \text{ or } \mathrm{d}(y)|\mathrm{d}(x) \\ 2, \text{ if } \frac{\mathrm{d}(x)}{\mathrm{d}(y)} = 2 \text{ or } \frac{d(y)}{d(x)} = 2 \\ 0, \text{ otherwise} \end{cases}.$$

The proof is by the method of contradiction. Assume that $DSG(L_4)$ has D3EL d with condition that "$|e_d(i) - e_d(j)| \leq 1$" $\forall\ 0 \leq i, j \leq 2$. Note that the number of lines labeled 0, 1, and 2 must be 6, respectively, in order to meet the desired "divisor 3-equitable property $|e_d(i) - e_d(j)| \leq 1$" $\forall\ 0 \leq i, j \leq 2$. But there are only 5 lines with 2, a contradiction (see Fig. 2). The same argument holds good for $DSG(L_n), n > 4$. Therefore, $DSG(L_n), n \geq 4$, does not admit D3EL.

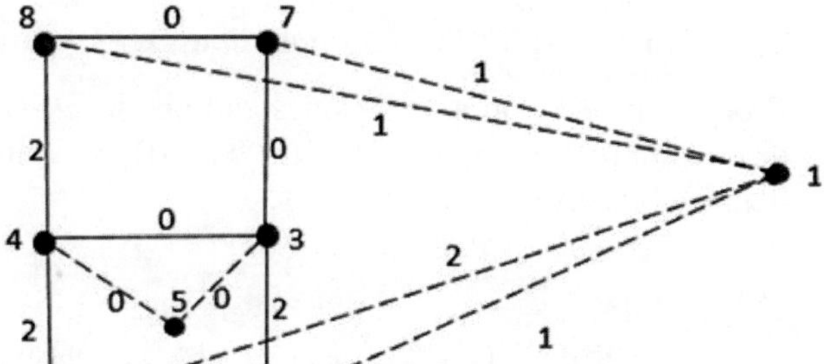

Fig. 1 D3EL of $DSG(L_3)$

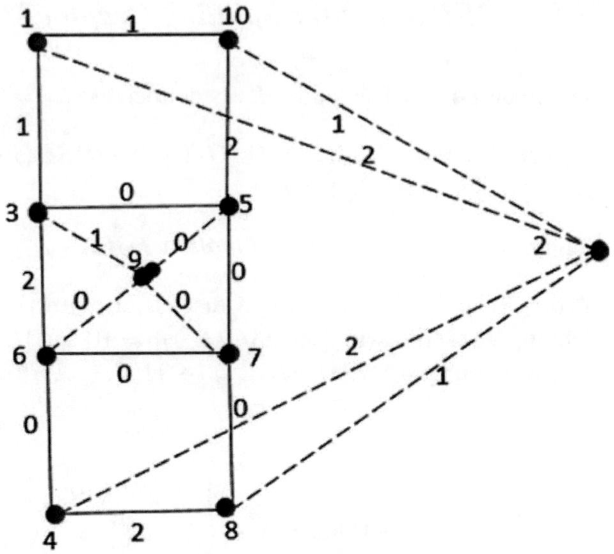

Fig. 2 Non-existence of D3EL for DSG(L_4)

2.2 *D3EL of Degree Splitting Graph of Triangular Snake Graph*

Definition 5 [6] "The triangular snake graph TS_n with n (odd) nodes is defined by starting with P_{n-1} and adding lines $(2i-1, 2i+1)$ for$i = 1, 2, ..., n-1$".

One can derive the D3EL of $DSG(TS_n)$, $n \leq 5$. One such example is given in Fig. 3.

Theorem 5 *$DSG(TS_n)$ does not allow D3EL* $\forall\, n(\text{odd}) \geq 7$.

Proof Take TS_n on $n \geq 7$ nodes and obtain $DSG(TS_n)$. Consider $n = 7$ for the discussion purpose and so $|V(DSG(TS_7))| = 9$ and $|E(DSG(TS_7))| = 16$. Now define a bijective

$$\text{map}\, d : V(DSG(TS_n)) \to \{1, 2, \ldots n\} \text{ as follows} : \mathrm{d}(e = xy) = \begin{cases} 1, & \text{if}\, d(x)|\mathrm{d}(y) \text{ or } \mathrm{d}(y)|\mathrm{d}(x) \\ 2, & \text{if } \frac{\mathrm{d}(x)}{\mathrm{d}(y)} = 2 \text{ or } \frac{d(y)}{d(x)} = 2 \\ 0, & \text{otherwise} \end{cases}.$$

The proof is by the method of contradiction. Assume that $DSG(TS_7)$ has D3EL d with condition that "$|e_d(i) - e_d(j)| \leq 1$" $\forall\, 0 \leq i, j \leq 2$. Note that the count of lines labeled with either 0 or 1 or 2 must be at least 5 and at most 6 to meet the desired "divisor 3-equitable property $|e_d(i) - e_d(j)| \leq 1$" $\forall\, 0 \leq i, j \leq 2$. But there are only four lines with label 2, a contradiction. Moreover, if $DSG(TS_7)$, does not admit D3EL, then $DSG(TS_n)$, $n > 7$, too does not admit D3EL (Fig. 4).

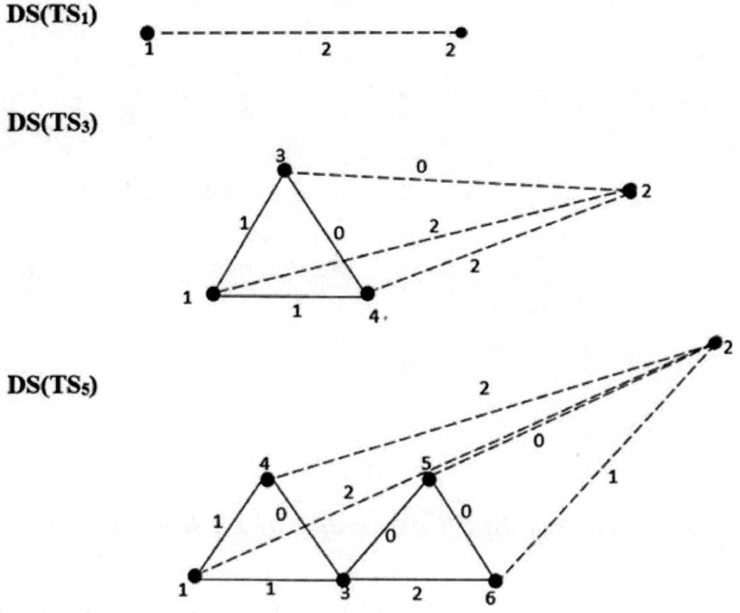

Fig. 3 D3EL of $DSG(TS_1)$, $DSG(TS_3)$, and $DSG(TS_5)$

Fig. 4 D3EL of $DSG(TS_7)$

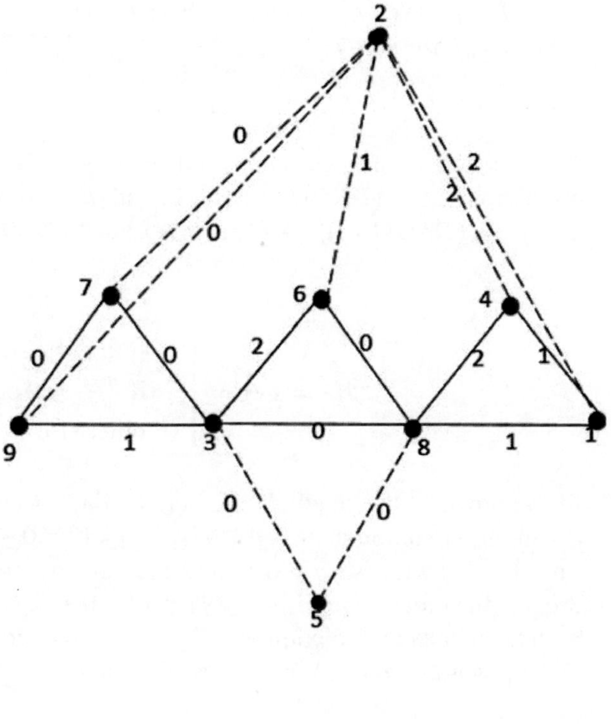

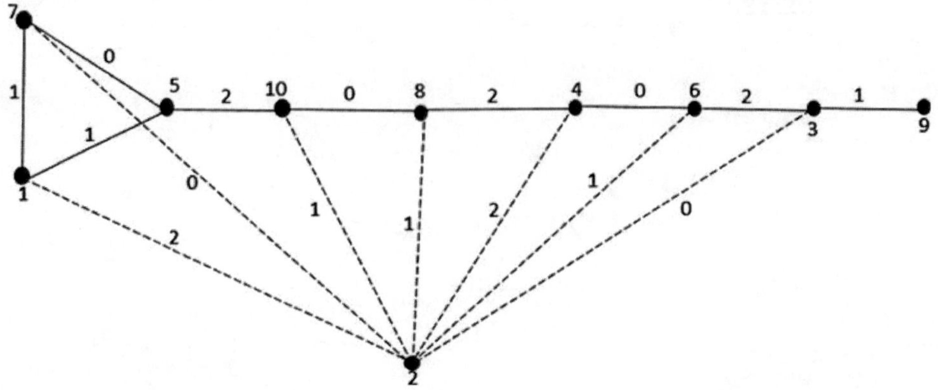

Fig. 5 D3EL of $DSG(L_{3,6})$

2.3 D3EL of Degree Splitting Graph of Lollipop Graph

Definition 6 [7] "The lollipop graph $L_{3,n}$ is a graph consisting of K_3 and P_n, connected with a bridge".

One can establish the D3EL of $DSG(L_{3,1})$, $DSG(L_{3,2})$, $DSG(L_{3,3})$, $DSG(L_{3,4})$, $DSG(L_{3,5})$, $DSG(L_{3,6})$. One such example is given in Fig. 5. So, we consider $DSG(L_{3,n})$, for $n \geq 7$.

Theorem 6 *$DSG(L_{3,n})$ does not accept D3EL $\forall\, n \geq 7$.*

Proof Take $L_{3,n}$ on $n \geq 7$ nodes and obtain $DS(L_{3,n})$. We take $n = 7$ for the "sake of discussion", so $|V(DSG(L_{3,7}))| = 11$ and $|E(DSG(L_{3,7}))| = 18$. Define a bijective map $d : V(DSG(L_{3,7})) \rightarrow \{1, 2, \ldots, n\}$ as follows:

$$\text{“d}(e = xy) = \begin{cases} 1, \text{ if } \mathrm{d}(x)|\mathrm{d}(y) \text{ or } \mathrm{d}(y)|\mathrm{d}(x) \\ 2, \text{ if } \frac{\mathrm{d}(x)}{\mathrm{d}(y)} = 2 \text{ or } \frac{\mathrm{d}(y)}{\mathrm{d}(x)} = 2 \\ 0, \text{ otherwise”} \end{cases}$$

We prove by the "method of contradiction". Assume that $DSG(L_{3,7})$ has D3EL d with the condition that "$|e_d(i) - e_d(j)| \leq 1$" $\forall\, 0 \leq i, j \leq 2$. Note that the count of lines labeled with either 0 or 1 or 2 must be exactly 6 to meet the desired "divisor 3-equitable property $|e_d(i) - e_d(j)| \leq 1$" $\forall\, 0 \leq i, j \leq 2$. But there are only five lines with label 2, a contradiction (see Fig. 6). Moreover, if $DSG(L_{3,7})$, does not permit D3EL, then $DSG(L_{3,n})$, $n > 7$, too does not permit D3EL.

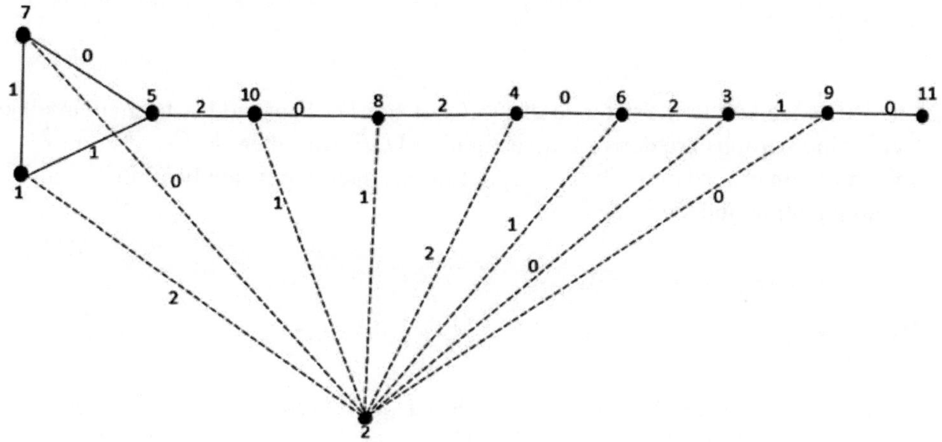

Fig. 6 Non-existence of D3EL of $DSG(L_{3,7})$

3 Applications

"The concept of graph labeling in graph theory also plays a vital role in computer science and communication networks. Due to the demand for high-quality multimedia service, the International Telecommunication Union (ITU) recently gave an integrated MSS system and hybrid satellite and terrestrial system to provide broadband service. In this system, the satellite radio technique needs to match the terrestrial wireless network as much as possible in order to reduce the cost. However, if the frequency reuse factor is 1, it might cause serious inter-beam interference (IBI) because of the usage of the same subcarrier between user equipment (UE) in adjacent beams. At the same time, the bandwidth assigned to MSS is very limited, so channel using efficiency is still a prime factor.

In channel borrowing, the acceptor cell which has no more unused nominal channels can borrow free channels from donor cells. For the above channel borrowing schemes, the tasks are majorly focused on which channel to borrow and the borrowing order. The channel borrowing technique in mobile satellite communication is given in [8] and it focused on the channel borrowing between different satellites. Moreover, the schemes listed above all obey a condition: the acceptor cell can only borrow channels that are not being used in the neighboring cells. A channel from the donor cell can be borrowed only if none of the cells belonging to the same group as the donor cell is using this channel. This might lead to a great reduction in borrowable channels. If not, it may lead to severe IBI. For more applications of graph theory in communication networks and satellite communication", see [8–10].

4 Conclusion

The existence and non-existence of D3EL of the DSG of ladder, triangular snake, and lollipop graphs are derived. Investigating D3EL of other classes of graphs is an open problem of research. The concept of D3EL may find its applications in satellite communication and space research.

References

1. Srivastav S, Gupta S (2019) D3EL of graphs. International Journal of Computer Science and Information Security (IJCSIS) 17(2):109–112
2. Ponraj R, Somasundaram S (2004) On the degree splitting graph of a graph, Nat Acad Sci Lett 27(7&8)
3. Tina JebiNivathitha K, Srinivasan N, Parthiban A, Sangeeta (2020) On D3EL of complete and star graphs. Journal of Xidian University 14(5):4448–4454
4. Tina JebiNivathitha K, Srinivasan, N, Parthiban A, Sangeeta (2020) On D3EL of wheel graphs. Journal of the Gujarat Research Society 8(5):5480–5483
5. Jeba Rani V, Sunder Raj S, Shyla Isaac Mary T (2019) Vertex polynomial of ladder graphs. Infokara Research 8(11):169–176
6. Agasthi P, Narvathi N (2018) On some labeling of triangular snake and central graph of triangular snake graph. In: National conference on mathematical techniques and its application (NCMTA), pp 1–19
7. Gallian JA (2011) A dynamic survey of graph labeling. The Electronic Journal of Combinatorics 18, #DS6
8. Guo L, Cui Q, Liu Y, Li X, Fu T, Chen Z (2013) Graph theory based channel reallocation technique in channel borrowing in mobile satellite communication. In: 2013, IEEE wireless communications and networking conference (WCNC), pp 2172–2177
9. Beanl DR, Engel GM (1992) Satellite communications and multigraph edge-coloring. Australasian Journal of Combinatorics 5:43–52
10. Kwok KF, Liu PCK, Li KC (1996) Channel borrowing techniques with reservation pool for LEO satellite land mobile communication systems. Satellite Systems for Mobile Communications and Navigation, pp 1129–132
11. West DB (2000) Introduction to graph theory, 2nd edn. Prentice Hall, Englewood Cliffs

A 100 Gbps Inter-Satellite Optical Wireless System (Is-OWC) Using PDM-SZCC Codes

Shippu Sachdeva and Manoj Sindhwani

Abstract In this research article, an economical optical code division multiplexed (OCDMA) inter-satellite system is presented using shift zero cross correlation codes (SZCC) and polarization division multiplexing (PDM). Total capacity of 100 Gbps is achieved with ten users and communicated over 25,000 km link distance between two satellites. Effect of different distances, pointing errors at receiver and transmitter is analysed and further proposed S ZCC codes are compared with random diagonal (RD) codes in Is-OWC system. It is observed that S ZCC codes are better than RD codes because of zero cross correlation.

Keywords Is-OWC · RD · S ZCC · PDM · SAC · OCDMA

1 Introduction

Because laser communication (LC) has a much larger bandwidth than broadband light mediums like light emitting diodes (LED), it is widely used in wireless communication and is known as wireless optical communication (WOC) [1]. Is-OWC, terrestrial optical wireless communication and indoor communication are the three basic varieties of WOC [2]. Due to several benefits such as security, speed, capacity and the lack of electromagnetic induction (EMI), the deployment of LC in space for satellite communication has superseded RF communication [3]. When it comes to constructing Is-OWC networks to obtain the highest data rate and transmission distance, researchers must consider modulation techniques [4]. In the literature, different studies are reported. Different modulation such as compressed spectrum return to zero (CSRZ), duo-binary RZ (DRZ) and modified DRZ (MDRZ) modulation in Is-OWS at 10, 20 and 40 Gbps were demonstrated in [5] for 1250 km and for higher bit rates, MDRZ was recommended.

Influence of pointing errors was studied in [6] with capacity 6*20 Gbps and wavelength division multiplexing (WDM) polarization interleaved (PI) technique

S. Sachdeva · M. Sindhwani (✉)
School of Electronics and Electrical Engineering, Lovely Professional University, Phagwara, Punjab, India
e-mail: manojsindhwani324@gmail.com

D. Singh et al. (eds.), *Computer Aided Constellation Management and Communication Satellites*, Lecture Notes in Electrical Engineering 987,
https://doi.org/10.1007/978-981-19-8555-3_13

over 1000 km. For the security and performance enhancement, optical code division multiplexing is prominent technique in free space optical systems. In the continuation to [5], three other modulations such as DPSK, chirped and alternate mark inversion (AMI) were investigated over 2500 km link at 10, 20 and 40 Gbps in [7]. AMI-based satellite system surpassed the performance of DPSK and chirped modulation at 40 Gbps. SAC-OCDMA allows several users to access the channel at the same time by sending distinct codes. Due to multi-user interference that results from the correlation of various user codes, this contributes to noise. As a result, multiple access interference (MAI) has the greatest impact on SAC-OCDMA performance, resulting in network shutdown. Code length increases with augmentations of subscribers which in turn waste the bandwidth.

Different SAC codes are investigated in the literature such as Hadmard codes, diagonal double weight, enhanced double weight, modified double weight, random diagonal codes (RD), etc. RD codes are investigated in the literature such as in [8, 9] due to better performance but, cross correlation increases with the increase in user. Is-OWC system with PDM and RD codes is studied in [10] and cover 25,000 km distance. BER of this system was high at 25,000 km due to cross correlation; however, BER can be lowered by using zero cross correlation SAC codes.

In this research article, an economical optical code division multiplexed inter-satellite system is presented using SZCC codes and polarization division multiplexing. Total capacity of 100 Gbps is achieved with ten users and communicated over 25,000 km link distance between two satellites. Effect of different distances, pointing errors at receiver and transmitter is analysed and further proposed SZCC codes are compared with random diagonal (RD) codes in Is-OWC system.

2 Code Construction of SZCC

In order to simplify the analysis, for three users, the code construction is first considered and the weight value is retained as two.

Step 1: Generation of combined matrix

There are two matrices needed to construct code, and we consider weight of code is 2. Identity matrix is primary matrix, and null matrix is null matrix with same dimensions. For 3 users, then matrix size is 3 × 3.

$$\begin{pmatrix} 1\,0\,1 \\ 0\,1\,0 \\ 1\,0\,1 \end{pmatrix} \text{ and } \begin{pmatrix} 0\,0\,0 \\ 0\,0\,0 \\ 0\,0\,0 \end{pmatrix}$$

Then, identity matrix and null matrix. Combined matrix is shown as

$$\begin{pmatrix} 1\,0\,0\,0\,0\,0 \\ 0\,1\,0\,0\,0\,0 \\ 0\,0\,1\,0\,0\,0 \end{pmatrix}$$

Step 2: Signal '1' in the combined matrix (S by = W × R – W – R + 1)

Where, row number is R, weight of the code is W, right signal operation of 1 s corresponding to row number. Operation on first row is first row by $= 2 \times 1 - 2 - 1 + 1 = 0$ (first row $R = 1$), on second row by $= 2 \times 2 - 2 - 2 + 1 = 1$ (middle row $R = 2$), on third row by $= 2 \times 3 - 2 - 3 + 1 = 2$ (last row $R = 3$). Where general formula for row $W \times R - W - R + 1$.

First row

$$1\,0\,0\,0\,0\,0 \rightarrow 1\,0\,0\,0\,0\,0$$

Middle row

$$0\,0\,1\,0\,0\,0 \rightarrow 0\,0\,0\,0\,1\,0$$

Last row

$$0\,1\,0\,0\,0\,0 \rightarrow 0\,0\,1\,0\,0\,0$$

Final matrix given as

$$\begin{pmatrix} 1\,0\,0\,0\,0\,0 \\ 0\,0\,1\,0\,0\,0 \\ 0\,0\,0\,0\,1\,0 \end{pmatrix}$$

Step 3: Replacing

Next, replace W-1 zeros in the right with 1 and in case of first row, 1 will be added at second position. In second row, 1 will be added fourth position and in last row, 1 will be added on sixth position.

First row replacing

$$1\,0\,0\,0\,0\,0 \rightarrow 1\,1\,0\,0\,0\,0$$

Middle row replacing

$$0\,0\,1\,0\,0\,0 \rightarrow 0\,0\,1\,1\,0\,0$$

Last row replacing

$$0\,0\,0\,0\,1\,0 \rightarrow 0\,0\,0\,0\,1\,1$$

Code matrix for final SZCC is obtained as

$$\begin{pmatrix} 1\,1\,0\,0\,0\,0 \\ 0\,0\,1\,1\,0\,0 \\ 0\,0\,0\,0\,1\,1 \end{pmatrix}$$

Code for $k = 5$, $w = 4$ is

$$\begin{pmatrix} 1\,1\,1\,1\,0\,0\,0\,0\,0\,0\,0\,0\,0\,0\,0\,0\,0\,0\,0\,0 \\ 0\,0\,0\,0\,1\,1\,1\,1\,0\,0\,0\,0\,0\,0\,0\,0\,0\,0\,0\,0 \\ 0\,0\,0\,0\,0\,0\,0\,0\,1\,1\,1\,1\,0\,0\,0\,0\,0\,0\,0\,0 \\ 0\,0\,0\,0\,0\,0\,0\,0\,0\,0\,0\,0\,1\,1\,1\,1\,0\,0\,0\,0 \\ 0\,0\,0\,0\,0\,0\,0\,0\,0\,0\,0\,0\,0\,0\,0\,0\,1\,1\,1\,1 \end{pmatrix}$$

3 System Setup

Optisystem software is considered for the proposed work, and Fig. 1 represents the SZCC-PDM-OCDMA system over Is-OWC. Lasers with input power 0 dBm are deployed for generating the optical carriers of different wavelengths, depending upon the SZCC code matrix. Total 40 wavelengths are required for the support of 10 users when weight is 4. But as polarization division multiplexing is used in the work, requirement of lasers reduces from 40 to 20 only.

Two different polarization, i.e. x and y are given to each laser carrier, so as to produce the polarization effects and which in turn reduces the laser requirements to half. Binary data each user is 10 Gbps and modulated with non-return to zero modulation format and x-polarization lasers are used for first five users. For next five users, lasers carrying polarization y state are employed in the system. All users then multiplexed and sent over Is-OWC link of 25,000 km with the use of 25 loops of loop control. Where each loop is consisting of two erbium doped fibre amplifier with gain 22 dB and noise figure 4 dB, Is-OWC channel length 1000 km.

Table 1 illustrates the proposed SZCC system. After transmission over Is-OWC channel, polarization multiplexer prior to the demultiplexer is placed to differentiate two polarization states. Pol. x is assigned to first five users, and pol. y is given to next five users. Pol. x and pol. y channels then given to different demultiplexer and signals are received with optical receiver which is having photo-detector avalanche, low pass filter for noise reduction, regenerator and BER analyser.

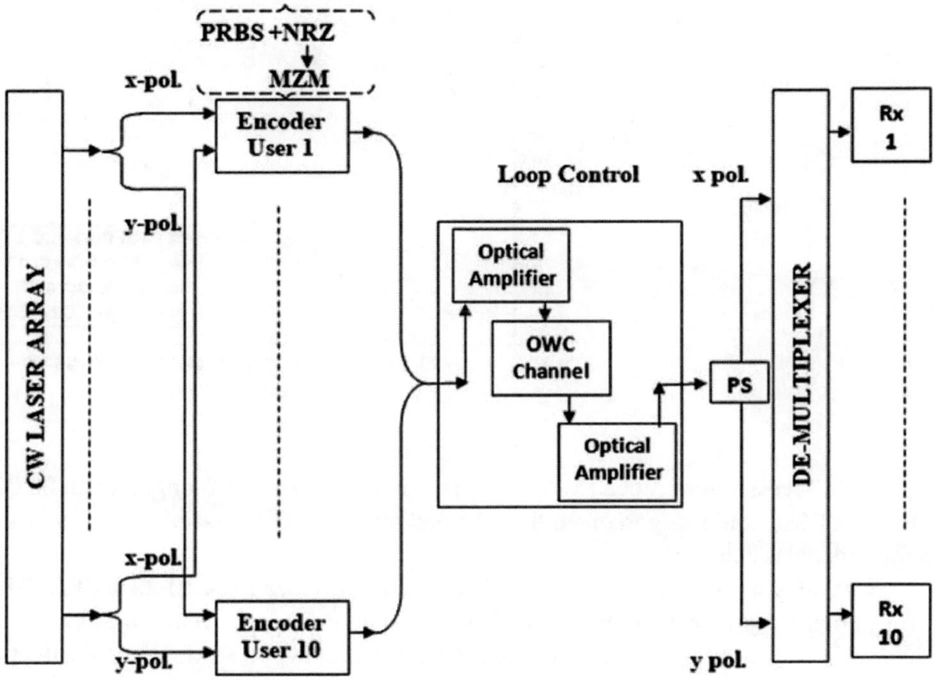

Fig. 1 Proposed SZCC-PDM-OCDMA system at 100 Gbps capacity

Table 1 Proposed SZCC system parameters

Parameters	Values
Data rate	10 Gbps/User
Users	10
Total capacity	100 Gbps
OCDMA code	SZCC
Modulation format	NRZ
Is-OWC distance	20,000–25,000 km
Optical amplifier	EDFA
Technique	OCDMA-PDM
Modulator	MZM
Photo-detector	APD

4 Results and Discussions

Proposed SZCC codes in the Is-OWC systems are investigated in this section at varied input parameters for BER and Q factor. Comparison of RD codes at first and tenth channels are compared SZCC codes for BER at varied link distances as shown in Fig. 2. BER increases with the distance increase due to scattering, attenuation,

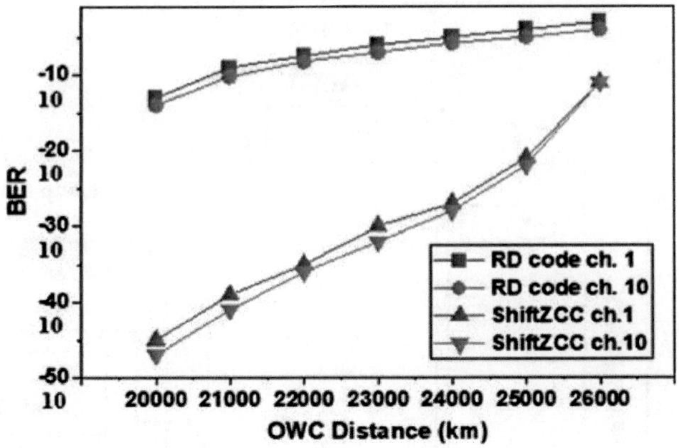

Fig. 2 Comparison of RD and SZCC codes for channel first and channel tenth using BER

multiple access interference (MAI) effects. Link distance is prolonged from 20,000 to 26,000 km, and it is perceived that channel ten, i.e. *y* polarization performs better and this polarization is better in both RD and SZCC codes.

RD codes show BER of 10-5, and SZCC code provides BER 10-22 at 25,000 km for *y* polarization, i.e. tenth channel. Bit reception is enhanced in case of SZCC codes than RD codes because of zero cross correlation and lesser MAI. Further, effect of distance increase between two satellites is observed for RD and SZCC codes having 10 users in terms of Q factor. Figure 3 represents the comparison of RD and SZCC codes for channel first and channel tenth. It is observed that channel first has degraded performance than channel tenth, but SZCC codes perform superior as compared to RD codes due to absence of cross correlation. RD codes show Q factor of 3.5, and SZCC code provides Q factor 9.54 at 25,000 km for y polarization.

Pointing of receiver and transmitter in Is-OWC systems plays an important role because of beacon signal, originates after synchronization of latitude and longitude of two satellites. Therefore, effects of receiver pointing errors on proposed system are analysed and it is evident that performance of OCDMA codes degrades with increase in pointing errors. Sensitivity of SZCC code towards receiving pointing

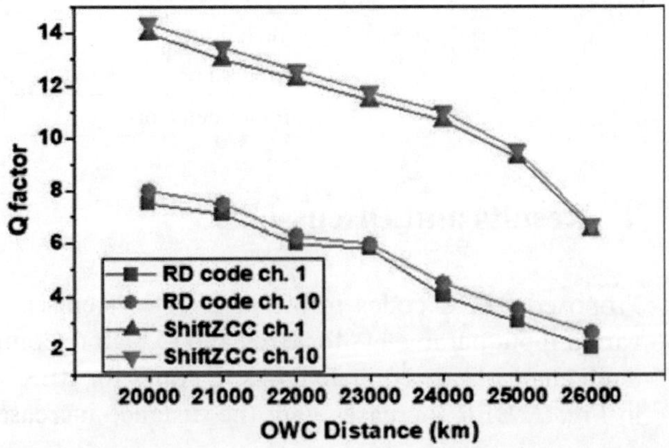

Fig. 3 Comparison of RD and SZCC codes for channel 1 and channel 10 using *Q* factor

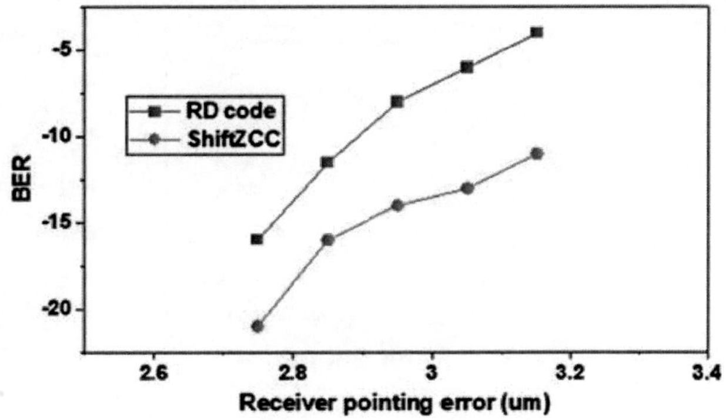

Fig. 4 Comparison of RD and SZCC codes at varied receiver pointing error

errors is lesser than RD codes, and therefore, BER is lesser in former code as shown in Fig. 4.

Figure 5 represents the variation of BER with the change in transmitting pointing errors, and it is seen that performance of OCDMA codes degrades with increase in transmitting pointing errors. At transmitting pointing error value 2.75 μm and 3.15 μm, BER observed is 10-16 and 10-4, respectively, for RD codes at channel 1 and channel 10. In case of SZCC codes, these values are 10-21 and 10-11 at channel 1 and channel 10. Sensitivity of SZCC code towards transmitting pointing errors is lesser than RD codes, and therefore, BER is lesser in SZCC code.

Eye diagram is end evaluation of signals which represents the eye height, *Q* factor, BER, jitter, received power, closer penalties, eye opening, RMS values, etc. In Fig. 5, eye diagrams for SZCC codes are presented at channel first and channel tenth at 25,000 km link distance. Results reveal that eye opening is wider in case of channel tenth, and therefore, y polarization users perform better.

Further, Fig. 6 depicts the eye diagrams at 6000 km for channel first and channel tenth at 6000 km in case of RD and SZCC codes. Figure 6a shows the eye diagram at 3.2 μm receiver pointing error channel 1, and Fig. 6b depicts the eye diagram at 3.2 μm receiver pointing error channel 10. Also Fig. 6c represents the eye diagram at

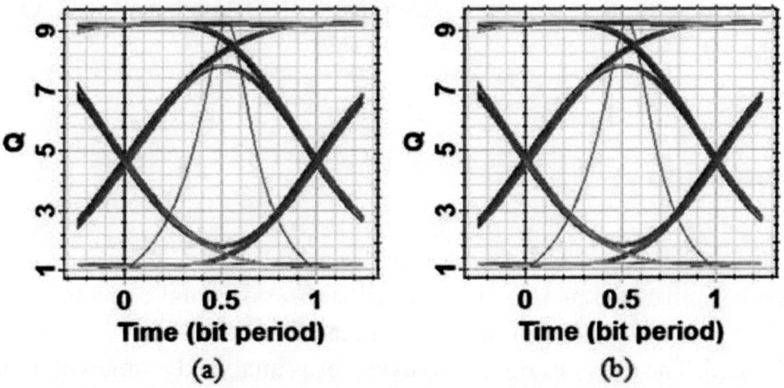

Fig. 5 Eye diagrams at 25,000 km using SZCC codes for **a** Channel 1 **b** Channel 10

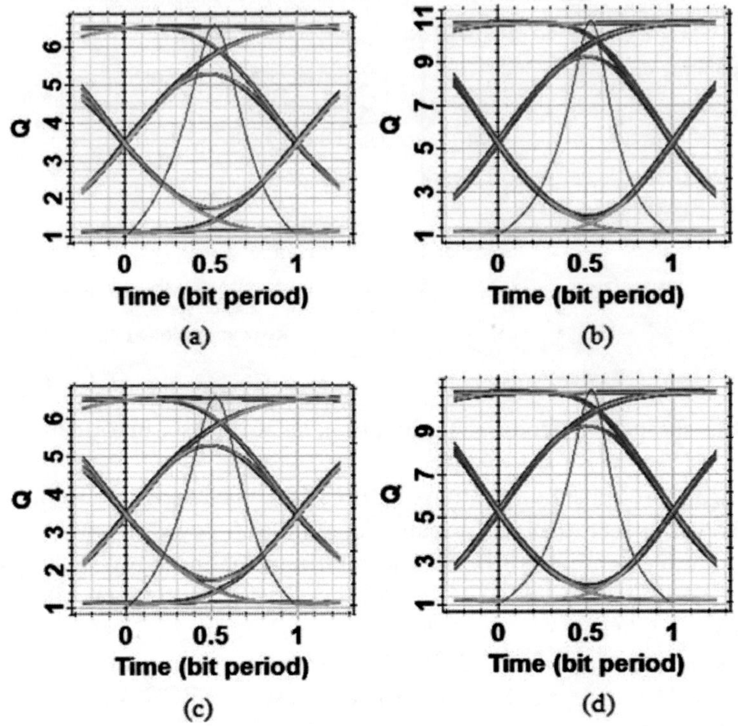

Fig. 6 Eye diagrams at 6000 km for **a** 3.2 μm receiver pointing error channel 1. **b** 3.2 μm receiver pointing error channel 10. **c** 3.2 μm transmitting pointing error channel 1. **d** 3.2 μm transmitting pointing error channel 10

3.2 μm transmitting pointing error at channel 1, and Fig. 6d shows the eye diagram at 3.2 μm transmitting pointing error at channel 10. It is clear from the results that eye opening is more in case of SZCC codes and for tenth channel.

RD codes in PDM-OCDMA inter-satellite optical wireless systems are reported in [10], and results are compared with proposed SZCC codes in Table 2. Proposed codes perform better than reported RD codes with distance achieved, Q factor, BER, etc.

5 Conclusion

In this work, an economical SAC-OCDMA-based Is-OWC system is presented using SZCC codes and PDM at 100 Gbps. With 10 users, a total capacity of 100 Gbps is obtained and over 26,000 km of communication distance between two satellites is accomplished. The effect of different distances is analysed along with pointing errors at the receiver as well as at transmitter. Further, proposed S ZCC codes are compared with RD codes in Is-OWC system for Q factor and BER. RD codes show BER of

Table 2 Comparison of proposed work and existing work

Parameter	Existing [10]	Proposed SZCC-PDM-Is-OWC
Total data rate	100 Gbps	100 Gbps
Users	10	10
Technique used	RD-PDM-Is-OWC	SZCC-PDM-Is-OWC
Cross correlation in OCDMA code	Yes	No
Total distance achieved	25,000 km	26,700 km
BER at 25,000 km	10^{-4}	10^{-22}

10-5, and SZCC code provides BER 10-22 at 25,000 km for y polarization, i.e. tenth channel. At transmitting pointing error value 2.75 μm and 3.15 μm, BER observed is 10-16 and 10-4, respectively, for RD codes at channel 1 and channel 10. In case of SZCC codes, these values are 10-21 and 10-11 at channel 1 and channel 10. Results revealed that sensitivity of SZCC code towards transmitting pointing errors is lesser than RD codes and therefore, BER is lesser in SZCC code. Therefore, SZCC codes due to absence of cross correlation and because lesser MAI performs far better than RD codes in PDM-OCDMA-Is-OWC systems.

References

1. Rashed ANZ, Tabbour MSF, Natarajan K (2019) Performance enhancement of overall LEO/MEO intersatellite optical wireless communication systems. Int J Satell Commun Network 38:31–40
2. Sivakumar P, Singh M, Malhotra J, Dhasarathan V (2020) Performance analysis of 160 Gbit/s single-channel PDM-QPSK based inter-satellite optical wireless communication (IsOWC) system. Wireless Netw. https://doi.org/10.1007/s11276-020-02287-2
3. Grover A, Sheetal A (2020) A 2 × 40 Gbps mode division multiplexing based inter-satellite optical wireless communication (IsOWC) system. Wireless Pers Commun. https://doi.org/10.1007/s11277-020-07483-z
4. Singh M, Malhotra J (2020) A high-speed long-haul wavelength division multiplexing–based inter-satellite optical wireless communication link using spectral-efficient 2-D orthogonal modulation scheme. Int J Commun Syst 33(6)
5. Alipour A, Mir A, Sheikhi A (2017) Ultra high capacity inter-satellite optical wireless communication system using different optimized modulation formats. Optik 127:19
6. Chaudhary S, Sharma A, Chaudhary N (2016) 6 × 20 Gbps hybrid WDM–PI inter-satellite system under the influence of transmitting pointing errors. J Opt Comm 37(4):375–379
7. Kaur R, Kaur H (2018) Comparative analysis of chirped, AMI and DPSK modulation techniques in IS-OWC system. Optik 154:755–762
8. Kaur R, Kaler RS (2020) Performance of zero cross correlation resultant weight spectral amplitude codes in lower Earth orbit-based optical wireless channel system. Int J Commun Syst 33(19)

9. Fadhil HA, Aljunid SA, Ahmad R (2009) Performance of random diagonal code for OCDMA systems using new spectral direct detection technique. Opt Fiber Technol 15:283–289
10. Chaudhary S, Tang X, Sharma A, Lin B, Wei X, Parmar A (2019) Cost-effective 100 Gbps SAC-OCDMA–PDM based inter-satellite communication link. Opt Quant Electron 51(148):1–10

Role of Satellite Communication in the Current Era

Nidhi Bansal Garg, Atul Garg, Mohit Bansal, Renu Popli, Rajeev Kumar, and Daljeet Singh

Abstract Internet is one of the great inventions for human kind. Everyone wants to be connected every time in this era. Real-time news or information is required for the growth of different sectors. Military, government, share-market, etc., require current information of the globe. And, all this is not possible without satellite communication. On the other side, dependency on it may be very harmful. If the frequency is blocked, then the whole system will be affected. As the new technologies may have many disadvantages, similarly satellite communications may create problems. The satellite communication has more applications and useful for the mankind and overall development of society, country, businesses, agriculture, education, health, etc.

Keywords Satellite · Satellite communication · Transponder · Mobile communication · Internet · Radio

1 Introduction

Communication is playing a significant role in the overall advancement of human beings. With the development of Internet, Web and smartphones, the communication method and pattern are changed. Now, the people are not bounded for the limited or traditional methodologies for communication [1]. After, the COVID-19, the role of electronic communication is increased; now the communication is not limited. The

N. Bansal Garg (✉) · A. Garg · R. Popli · R. Kumar
Chitkara University Institute of Engineering and Technology, Chitkara University, Rajpura, Punjab, India
e-mail: garg.nidhi@chitkara.edu.in

M. Bansal
ABES Engineering College, Uttar Pradesh, Campus 1, Ghaziabad 201009, India

D. Singh
Center for Space Research, Division of Research and Development, Lovely Professional University, Phagwara, Punjab, India

D. Singh et al. (eds.), *Computer Aided Constellation Management and Communication Satellites*, Lecture Notes in Electrical Engineering 987,
https://doi.org/10.1007/978-981-19-8555-3_14

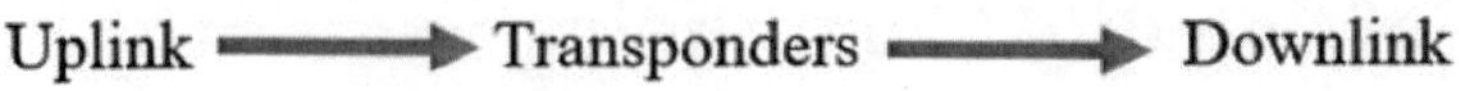

Fig. 1 Working of satellite communication

education system is effected after the COVID-19 and their teaching and evaluation also depend on the electronic media [2, 3].

The wireless communication using artificial satellite can be known as satellite communication. Different type of services, e.g., television, voice or video calling, Internet, radio, etc., are provided through satellite communication. In the working method, an artificial satellite is placed on globe between the two more common communication points [4–6].

Broadcast communication, global coverage, mobility, bandwidth facilities, and Internet services made the satellite communication different from other services. Apart from these multiple services, communication services in remote areas for forest or hill areas, aircraft, maritime satellite, crowded areas satellite communication are the best way of fast communication. High-quality networks like broadband, heterogeneous or simple networks can be served with satellite network [7].

Radio waves are used for satellite communication, big antennas on the earth receives the signals from the satellite and transmit these signals further. Similar to the mirrors, satellite receives the signal like radio, Internet data, etc., from earth and bouncing back on the other side on the earth [8–10]. There are three main stages in satellite communication (Fig. 1).

Uplink. In this stage, the information from earth to satellite is sent to transmit the information. For an example, if users want to send TV signals from one face of earth to another face of earth. Then, they need to send signals from first face to satellite.

Transponder. It works like transmitters/amplifiers/radio receivers. Transponders boost the incoming signal and modify frequency so maintaining the quality of signals.

Downlinks. As the satellite works as mirror, the information is sent through uplink then satellite sends back the signals on other side on globe. For that, antennas are set as receivers on the earth. The working of satellite communication is depicted in Fig. 2 [8–10].

Federal Communications Commission (FCC) initiated the satellite communication by listing some launches of International Telecommunications Satellite Organization (INTELSAT) in 1995 which has currently become an association of more than 130 countries. India is also working rigorously to be the world leader in satellite communication and has developed one of the largest domestic communication satellite structures for Asia–Pacific called Indian National Satellite (INSAT) containing fifteen satellites. It works majorly in the C and Ku-bands [9].

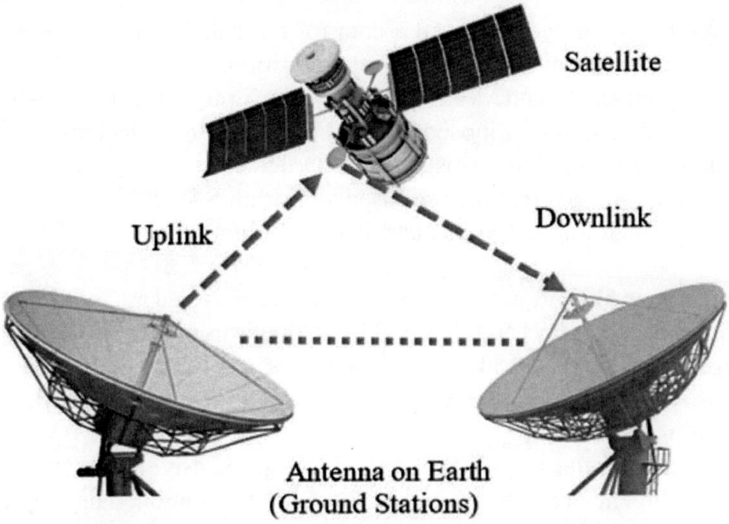

Fig. 2 Satellite communication

2 Literature Survey

The overall growth in any field without electronic media is not possible in the current era [11]. As the use of satellite is increasing in every field, so new requirement and expected possibilities are also increased. In this section, the research work done by renowned researchers in different areas of satellite communication is described:

Authors in [3] focused on the use of Artificial Intelligence (AI) in satellite communication. In their study, different applications of AI are discussed which can be used for communication satellite, e.g., network traffic forecast, beam-hopping, channel modeling, etc. The authors in [6] focused on real users, traffic, both synthetic satellite operational networks used for satellite Internet access and proposed the qualitative results of the basis of real measurements.

According to authors [6], this work is done first time in the literature. In their work, they showed that the performance of new satellite network is much higher than the previous network solutions. The authors in [12] focused on contribution of SatCom. Authors also discussed advanced concepts and future challenges in SatCom.

Authors in [13] discussed various new applications like satellite communication based on laser beams, space situational awareness, reuse of frequency, concept of spot beam, etc., and challenges in the satellite communication. According to them, as the technology is improving and the mode of communication is developing the use and challenges, e.g., high-powered platforms, critical future technologies, new policies issues [9].

A hybrid architecture of LPWAN-satellite communication network for IoT and protocols is proposed in [14] to implement IoT system in isolated zones. To optimize data transfer, authors proposed changes in the presentation layer. Focusing on data transfer optimization, we proposed a data format change on the presentation layer that can be implemented using packaging algorithm.

Simulated annealing and Monte Carlo based on uplink transmission forecast algorithm for IoT applications proposed in [15]. For comparing objectivity and throughput of various scheduling algorithms, Second-Order Deviation (SOD) metric is used. And, the results showed that the proposed algorithm SA-MC is better than Maximal Throughput scheduling (MAxTh) and Round Robin Scheduling (RRS) algorithms [15].

The authors in [13] proposed a novel architecture named as Coordinated Satellite-Terrestrial Networks (CSTNs). In CSTN, satellite is used for handling broadcast and to manage multiple nodes and to cover the nodes afar from satellite reach, terrestrial network is used. The results showed that in case of large size network, efficiency of blockchain increased [16].

Fourati and Alouini in [17] presented a comprehensive literature survey on the artificial intelligence and its use in satellite communication. The major challenges of satellite communication and their problems are also given in [17]. Alam et al. in [18] have proposed a novel antenna design for CubeSat communication using metamaterial-based patch antenna which consists of two separate layers working in the frequency band of 443.5–455 MHz. The size of the antenna proposed in [18] is $80 \times 40 \times 3.35$ mm^3.

3 Types of Satellite Communication Services

Following are the diverse satellite communications services as proposed by the International Telecommunication Union (ITU) [12] (Table 1):

Table 1 Services of satellite communications

Fixed satellite	Mobile satellite	Broadcasting satellite
In fixed satellite, as the name suggests, the communication application is employed in between the satellite and a ground station having a fixed location	Contrary to fixed satellite, the mobile satellites work for moving earth stations/users	For the broadcasting satellites, the signal transmitted by such a satellite is proposed to be received by all the receivers in its receiving range
Power signals are very low	Data communication and two-way voice calls are possible	One-way communication
This type of services provide links for networks like telephone and also works as a transmitter for TV signals	Connecting ships, airplanes at remote places	Television broadcasting, radio, etc., are the applications
Large antennas are used to receive signals	Small antenna is required	Parabolic antenna is used to receive the signals

4 Applications of Satellites

The technology is upgrading day by day. The world seems to have shrunk with the use of different communication methods. Also, as the need and uses of new technologies are increasing, the need of satellites for communications is also increasing, and with the broad benefits of satellites, the applications of satellites are also increasing. Following are the few common applications of satellites [10, 13, 19]:

Weather Forecasting. Forecasting and predictions are more important in the developing world. Weather forecasting is more important for agriculture-dominated countries. Weather forecasting is not only helpful for the farmers but it is equally helpful for the government, society, industry, health, etc. For example, the INSAT-3A, KALPANA-1 and INSAT-3C are some of the exemplary missions used for weather forecasting.

Military. The real-time information of any geographical area is important for military of any country. And, with the help of new technologies, it became easier for army to know about the current location or geographical condition of enemies. Also, they can set the targets and positions of missiles from their own country to their enemies. For example, the early warning system under the Defense Support Program (DSP) of the USA is utilized for only military purposes.

Radio and TV Broadcast. New/information broadcast among country or on remote areas is one of the big achievements in past development with satellite. TV and radios not only helpful to broadcast the information of news but these are also useful for entertainment for long time. INSAT 3E mission is one such example of TV broadcasting satellites.

Internet Access. Life without Internet is cannot be imagined in the current era. Everyone wants to be online. Fast communication, real-time information.

Connecting Remote Areas. Geographical and remote areas are big obstacles in the growth of country, society and mankind. Countries such as India whose geography is not only big but also different. Like where there is a mountain, somewhere a desert, forest, river, rocky place and field. Following are the challenges in remote areas:

- To send or receive messages
- To educate people
- To learn about their culture
- To reach on the right place

All such problems can be solved with the help of satellite communications. And, the government of such countries is working in this direction.

Telephone/mobile communication on Globe. All the globe is connected with each other with telephonic connection. And, this is also one of the services of satellite.

Transportation. Transport facilities like aircraft, marine, ship, Trans, buses, cars and radio are in the moving form. For such kind of facilities, navigational facility is required for communication. Navigational satellites are playing a vital role for such kind of facilities.

5 Impact of Satellite Communication in the Current Era

Satellites have changed the entire way of communication. The way, how the people transmit and receive messages has evolved. As a result of satellite communication, the speed of communication became very fast which directly or indirectly impacted on government, education, society in the current era. More and better communication services are available, e.g., telephone, Internet, video call, etc. More interactive with high-resolution data transfer and processing has become possible. The fast communication system provides the better facilities to understand about the other countries, states and people on the globe, which also effect on the daily life, economic, politics, culture and society. The world seems to be shrinking with new technologies and fast communication. These communication systems are beneficial for the growth of the society though it is affecting the era in negatively and positively.

6 Conclusion

The use of satellite communication is increased in the current era. Though, it is very costly to set and install satellite. But, the cost does not matter in this fast world. Every country wants to be connected at its all remote areas. In this work, a detailed overview of satellite communication is presented. The advantages and drawbacks of satellite communication along with various missions and applications are presented. In the countries like India, where the geographical area is large with different diversities, communication became more critical without use of satellite communications. This is the time of technology, and it is growing very fast. And, use of satellite communication is increasing day by day. It is very difficult to handle different situations without satellite communication.

References

1. Yan X, An K, Liang T, Zheng G, Ding Z, Chatzinotas S, Liu Y (2019) The application of power-domain non-orthogonal multiple access in satellite communication networks. IEEE Access 7:63531–63539
2. Kumar G, Singh D, Kumar R (2021) A planar CPW fed UWB antenna with dual rectangular notch band characteristics incorporating U-slot, SRRs, and EBGs. Int J RF Microwave Comput Aided Eng 31(7):e22676
3. Maraqa O, Rajasekaran AS, Al-Ahmadi S, Yanikomeroglu H, Sait SM (2020) A survey of rate-optimal power domain NOMA with enabling technologies of future wireless networks. IEEE Commun Surv Tutorials 22(4):2192–2235
4. Kumar R, Saini GS, Singh D (2020) Compact tri-band patch antenna for Ku band applications. PIER C 103:45–58
5. Kuang L, Chen X, Jiang C, Zhang H, Wu S (2017) Radio resource management in future terrestrial-satellite communication networks. IEEE Wirel Commun 24(5):81–87
6. Botta A, Pescapé A (2014) On the performance of new generation satellite broadband internet services. IEEE Commun Mag 52(6):202–209
7. Hu Y, Li VO (2001) Satellite-based internet: a tutorial. IEEE Commun Mag 39(3):154–162
8. Abdu TS, Kisseleff S, Lagunas E, Chatzinotas S (2021) Flexible resource optimization for GEO multibeam satellite communication system. IEEE Trans Wireless Commun 20(12):7888–7902
9. Shah S, Siddharth M, Vishwakarma N, Swaminathan R, Madhukumar AS (2021) Adaptive-combining-based hybrid FSO/RF satellite communication with and without HAPS. IEEE Access 9:81492–81511
10. Wang C, Zhang Z, Wu J, Chen C, Gao F (2021) An overview of protected satellite communications in intelligent age. Science China Inf Sci 64(6):1–18
11. Garg A, Popli R, Sarao BS (2021) Growth of digitization and its impact on big data analytics. In: IOP conference series: materials science and engineering, vol 1022, no. 1. IOP Publishing, p 012083
12. Kodheli O, Lagunas E, Maturo N, Sharma SK, Shankar B, Montoya JFM et al (2020) Satellite communications in the new space era: a survey and future challenges. IEEE Commun Surv Tutorials 23(1):70–109
13. Misra D, Misra DK, Tripathi SP (2013) Satellite communication advancement, issues, challenges and applications. International Journal of Advanced Research in Computer and Communication Engineering 2(4):1681–1686
14. Lysogor I, Voskov L, Rolich A, Efremov S (2019) Study of data transfer in a heterogeneous Lora-satellite network for the internet of remote things. Sensors 19(15):3384
15. Ji Y, Kumar R, Singh D, Singh M (2021) Performance analysis of target information recognition system for agricultural robots. International Journal of Agricultural and Environmental Information Systems (IJAEIS) 12(2):49–60
16. Wei H, Feng W, Zhang C, Chen Y, Fang Y, Ge N (2020) Creating efficient blockchains for the Internet of Things by coordinated satellite-terrestrial networks. IEEE Wirel Commun 27(3):104–110
17. Fourati F, Alouini MS (2021) Artificial intelligence for satellite communication: a review. Intelligent and Converged Networks 2(3):213–243
18. Alam T, Almutairi AF, Samsuzzaman M, Cho M, Islam MT (2021) Metamaterial array based meander line planar antenna for cube satellite communication. Sci Rep 11(1):1–12
19. Peng D, Bandi A, Li Y, Chatzinotas S, Ottersten B (2021) Hybrid beamforming, user scheduling, and resource allocation for integrated terrestrial-satellite communication. IEEE Trans Veh Technol 70(9):8868–8882

Triple Band H-Shaped Dielectric Resonator Antenna for S and C Band Satellite Communication

Dheeraj Kumar, Shekhar Yadav, Komal Jaiswal, and Narbada Prasad Gupta

Abstract In the present paper, a low-profile and compact H-shaped dielectric resonator antenna (DRA) is investigated. The antenna is fed using coaxial cable, which provides the required input matching and improves the radiation pattern. The proposed design engenders three frequency bands, i.e., 3.31–3.77 GHz, 4.33–5.10 GHz, and 5.17–5.46 GHz with a resonant frequency of 3.5, 4.6, and 5.3 GHz, respectively. The antenna shows an average gain of 5.4, 5.9, and 6.3 dBi in the three desired bands. As a result, proposed DRA may find useful applications in the S and C bands used by airport surveillance radar for air traffic control, weather radar, surface ship radar, satellite communication, Wi-Fi devices, cordless telephones, and some surveillance and weather radar systems.

Keywords Dielectric resonator antenna · Triple band · S and C band · Satellite communication

1 Introduction

An antenna is the backbone of modern communication systems, and nowadays, research has been done to design multi-frequency antennas for modern wireless communication, which can work effectively for different wireless applications simultaneously [1]. The patch antenna radiated through a 2D surface or narrow slot, whereas the DRA radiates through the entire 3D surface, its journey started as an energy storage device in 1939. After some decades in 1983, it was used as an energy radiator because of its low loss and high bandwidth. Due to its small size, it is one of the major low-profile antennas to meet the requirements of communication system. Due to its small size, high dielectric constant material can be used for the

D. Kumar (✉)
Department of Physics and Electronics, Rajdhani College, University of Delhi, Delhi, India

S. Yadav · K. Jaiswal
Department of Electronics and Communication, University of Allahabad, Prayagraj, India

N. P. Gupta
Jai Narain College of Technology, Bhopal, India

D. Singh et al. (eds.), *Computer Aided Constellation Management and Communication Satellites*, Lecture Notes in Electrical Engineering 987,
https://doi.org/10.1007/978-981-19-8555-3_15

design of small size DRA [2]. The dielectric constant and quality factor of DRA are high, and the size is inversely proportional to the relative permittivity of the material forming structure. It has a flexible feed arrangement, a simple geometric structure, and low metallic loss. It is usually mounted on a ground plane, has high radiation efficiency, a high degree of freedom and versatility [3–8]. Modern communication systems require compact and wideband antennas. These requirements can be met by using perforations and edge grounding techniques [9–11]. In this paper, a compact triple band, H-shaped DRA based on perforations is investigated. Initially, perforations are etched/drilled/removed on the dielectric material to make four H shapes to lower the quality factor, which improves the bandwidth. Next, the geometry is made compact by placing the perforated DRA on the ground plane of copper without using substrate [2]. The proposed antenna is simulated analysis by using CST microwave studio software.

A rectangular DRA is shown in Fig. 1. The rectangular shape of DRA avoids mode degeneracy, whereas in the case of cylindrical and spherical structures, mode degeneracy exists. Rectangular DRA provides more flexibility in terms of bandwidth control and resonance frequency as each dimension can be chosen independently, which also promotes avoiding modes. Due to mode degeneracy, the cross-polarization level of the antenna increases, which causes limitation in the performance of the antenna [11].

Modes of DRA can be confined or non-confined according to the arbitrarily shape of the DRA with high permittivity [12–14]. The following conditions are satisfied at all resonator surfaces:

$$E \cdot n = 0 \tag{1}$$

$$n \times H = 0 \tag{2}$$

A mathematical equation of resonant frequency is derived using the transcendental equation. Hence, the resonant frequency calculations are based on the waveguide

Fig. 1 Geometries of rectangular DRA

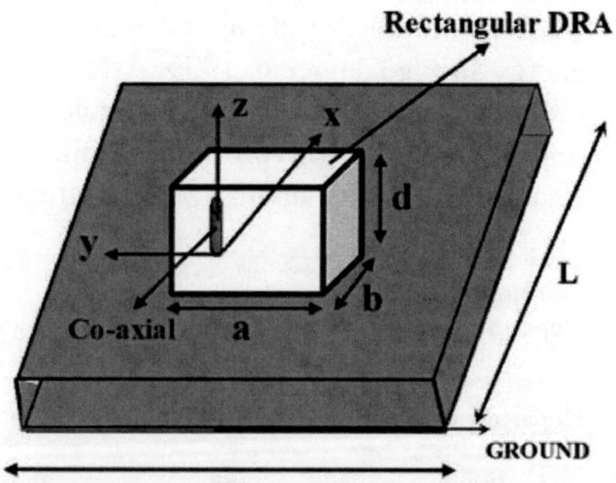

structure with a perfect electric conductor at $z = 0$, a perfect magnetic conductor at $x = \pm a/2$ and $z = d$. Boundary conditions of the continuity of electric and magnetic field are applied at $y = b/2$. Equations for the transverse component of E and H-field inside rectangular DRA in the terms of longitudinal field are as follows [14–16].

$$E_x = \frac{1}{j\omega\varepsilon\left(1 + \frac{y^2}{k^2}\right)}\left[\frac{\partial H_z}{\partial y} - \frac{1}{j\omega\mu}\frac{\partial^2 E_z}{\partial z \partial x}\right]$$

$$E_y = \frac{1}{j\omega\varepsilon\left(1 + \frac{y^2}{k^2}\right)}\left[\frac{1}{j\omega\mu}\frac{\partial^2 E_z}{\partial z \partial y} - \frac{\partial H_z}{\partial x}\right]$$

$$H_x = \frac{1}{j\omega\mu\left(1 + \frac{y^2}{k^2}\right)}\left[\frac{\partial E_z}{\partial y} - \frac{1}{j\omega\mu}\frac{\partial^2 H_z}{\partial z \partial x}\right]$$

$$H_y = \frac{1}{j\omega\mu\left(1 + \frac{y^2}{k^2}\right)}\left[\frac{1}{j\omega\mu}\frac{\partial^2 H_z}{\partial z \partial y} - \frac{\partial E_z}{\partial x}\right]$$

Figure 2 illustrates the rectangular DRA placed on an infinite ground plane with a dimension of $(a \times b \times d)$ mm^3. Resonances can occur at the following frequencies [12]

$$\mathrm{fr} = \frac{c}{2\pi\sqrt{\varepsilon_r}}\sqrt{\left(\frac{m\pi}{a}\right)^2 + \left(\frac{n\pi}{b}\right)^2 + \left(\frac{p\pi}{d}\right)^2}$$

2 Antenna Geometry

The H-shaped DRA is fed by coaxial cable with perfect impedance matching (Fig. 1). In probe feeding, DRA requires a higher value of ε_r and a shorter length of probe [14, 15]. The dimensions of the H-shaped DRA are 26 × 42 × 11 mm^3 based on Roger RT 6010 LM of dielectric constant 10.7 and the ground plane is made of copper with dimensions of 66 × 82 × 2 mm^3. Four iterations of the H shape are made by perforation, i.e., dimensions of perforations are 8.66 × 8.66 × 11 mm^3, 5.33 × 5.33 × 11 mm^3, 3.33 × 3.33 × 11 mm^3, and 2 × 2 × 11 mm^3, which etched/drilled/removed in shape. The geometry of DRA is shown in Fig. 2a–c (Table 1).

Fig. 2 Principal figure of the proposed DRA **a** Front view, **b** Ground view, **c** Side view

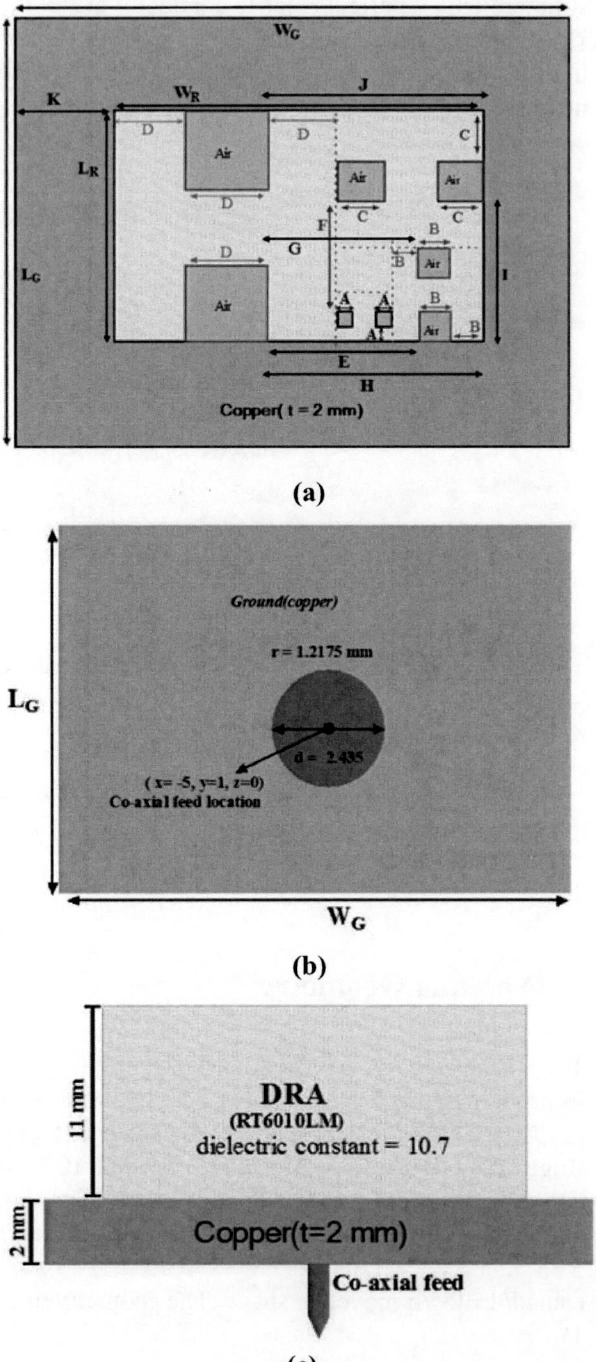

Table 1 Dimensions of the proposed antenna

Parameters	Dimensions (mm)
W_G	66
L_G	82
W_R	41.98
L_R	26
A	2
B	3.333
C	5.333
D	8.666
E	17.99
I	15.32
J	24.65
K	20

3 Results and Discussion

The characteristics of the proposed H-shaped DRA antenna are analyzed in terms of S11, gain, and radiation pattern. Figure 3 shows that three resonances at 3.5, 4.6, and 5.3 GHz. The lower band has a range from 3.31 to 3.77 GHz (12.99%), the middle band from 4.33 to 5.10 GHz (16.33%), and the upper band from 5.17 to 5.46 GHz (5.47%). The insertion loss and gain as a function of frequency are shown in Figs. 3 and 4. The values of gain at three resonant frequencies are 5.4 dBi, 5.9 dBi, and 6.3 dBi, respectively. The radiation patterns of an antenna are simulated in the x–z plane (E-plane) and y–z plane (H-plane), Both the E-plane and the H-plane are shown in Fig. 5. The radiation patterns of DRA are broadside and symmetrical at all three resonance frequencies. It is also observed that the radiation patterns in the E-plane and H-plane are directional in nature, and there is a slight difference in radiation patterns between the lower and middle resonance bands. The radiation characteristics of the proposed antenna show that the DRA is suitable for S and C band applications.

4 Conclusions

A triple band dielectric resonator antenna design based on perforations for wireless communications, covering both the S (2–4 GHz) and the C (4–8 GHz) frequency bands, has been presented. The perforations lower the quality factor, hence, improve the bandwidth. It is observed that antenna gain is above 5.4 dBi within the working bands and the radiation pattern of the antenna in all frequency bands and all planes is directional. Thus, the proposed DRA is appropriate for wireless communication

Fig. 3 S_{11} versus frequency plot

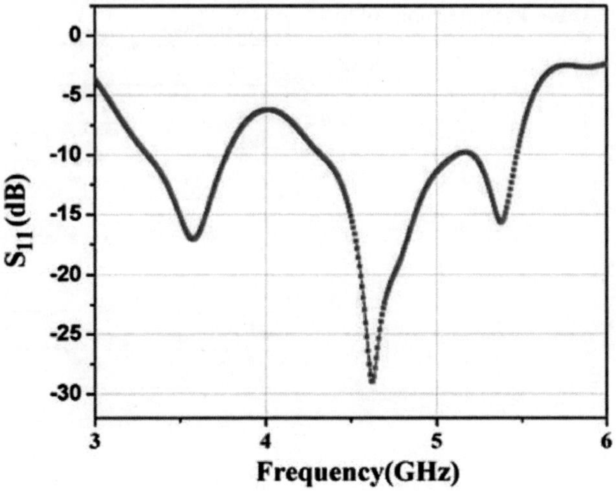

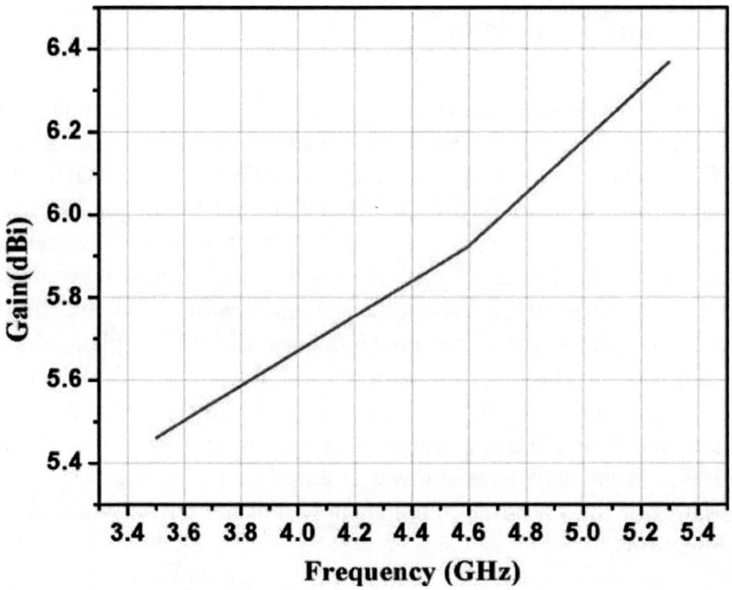

Fig. 4 Gain versus frequency plot

devices in the S and C frequency bands used for weather radar, surface ship radar, communications satellites, especially those of NASA for communication with ISS and Space Shuttle satellite communications, for full-time satellite TV networks or raw satellite feeds.

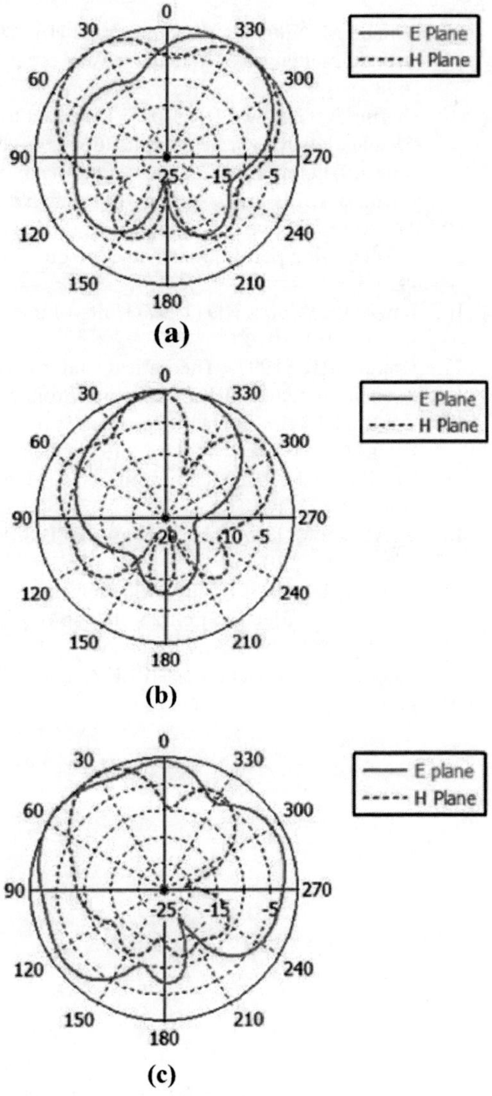

Fig. 5 Radiation pattern in *E*-plane and *H*-plane at **a** 3.5 GHz and **b** 4.6 GHz, **c** 5.3 GHz

References

1. Petosa (2007) Dielectric resonator antenna handbook. Artech House, Norwood
2. Keyrouz1 S, Caratelli D (2016) Dielectric resonator antennas: basic concepts, design guidelines, and recent developments at millimeter-wave frequencies. Int J Antennas Propag 1–20
3. Alkanhal MAS (2009) Composite compact triple-band microstrip antennas. Prog Electromagnet Res 93:221–236
4. Wu YJ, Sun BH, Li JF, Liu QZ (2007) Triple-band omni-directional antenna for WLAN application. Prog Electromagnet Res 76:477–484
5. Shi L, Sun H, Dong W, Lv X (2009) A dual-band multifunction carborne hybrid antenna for satellite communication relay system. Prog Electromagnet Res 95:329–340

6. Bemani M, Nikmehr V, Younesiraad HA (2011) novel small triple band rectangular dielectric resonator antenna for Wlan and Wimax applications. J Electromagnet Waves Appl 25:1688–1698
7. Sharma A, Gangwar RK (2015) Compact triband cylindrical dielectric resonator antenna with circular slots for wireless application. J Electromagnet Waves Appl 1–10
8. Fang XS, Leung KW (2011) Designs of single, dual, wide-band rectangular dielectric resonator antennas. IEEE Trans Antennas Propag 59(2):2409–2414
9. Patel P, Mukherjee B, Mukherjee JA (2015) Compact wideband rectangular dielectric resonator antenna using perforations and edge grounding. IEEE Antenna Wireless Propag Lett 14:490–493
10. Tam MTK, Murch RD (1997) Half volume dielectric resonator antenna designs. Electron Lett 33(23):1914–1916
11. Mongia RK (1997) Theoretical and experimental investigations on rectangular dielectric resonator antennas. IEEE Antennas Propag 45(9):1348–1356
12. Van Bladel J (1975) On the resonances of a dielectric resonator of very high permittivity. IEEE Trans Microw Theory Tech 23:199–208
13. Van Bladel J (1975) The excitation of dielectric resonators of very high permittivity. IEEE Trans Microwave Theory Tech 23:208–215
14. McAllister M, Long SA, Conway GL (1983) Rectangular dielectric resonator antenna. Electron Lett 19:219–220
15. Mongia RK (1989) Half-split dielectric resonator placed on a metallic plane for antenna applications. Electron Lett 25:462–464
16. Junker GP, Kishk AA, Glisson AW (1994) Input impedance of dielectric resonator antennas excited by a coaxial probe. IEEE Antennas Propag 42(9):960–966

Antenna Design Considerations for Satellite Communication: A Review

Aarti Bansal, Shivani Malhotra, Sandeep Singla, and Harsimranjit Kaur

Abstract Antenna design is a critical component of any communication system and requires to exhibit multiband operation with optimal radiation characteristics to enhance technological capabilities. Recently, CubeSat, low earth orbit satellites have emerged as a new class of satellites due to their low cost and light weight, which enable them to perform a variety of functions such as remote sensing, deep space communication research, supports mutiple mobile users and so on. Antenna design for these CubeSat applications must be compact in size while also exhibiting high gain and wide bandwidth characteristics. This article discusses eight latest antenna designs analysing various gain and bandwidth enhancement techniques to offer a comprehensive reference model for improved quality of life. This will aid and deliberate on identification and selection of relevant design specifically for satellite communication.

Keywords Satellite communication bands · CubeSat · Microstrip patch antenna · Metamaterial · Superstrate

1 Introduction

Satellite communication (SatCom) is an important factor in establishing a country's defence status. It has a wide range of applications in fields like as oceanography,

A. Bansal · S. Malhotra (✉) · H. Kaur
Chitkara University Institute of Engineering and Technology, Chitkara University, Rajpura, Punjab 140401, India
e-mail: Shivani.malhotra@chitkara.edu.in

A. Bansal
e-mail: aarti.bansal@chitkara.edu.in

H. Kaur
e-mail: harsimran.kaur@chitkara.edu.in

S. Singla
RIMT University, Mandi Gobindgarh, Punjab 147301, India
e-mail: sandeep.singla@rimt.ac.in

D. Singh et al. (eds.), *Computer Aided Constellation Management and Communication Satellites*, Lecture Notes in Electrical Engineering 987,
https://doi.org/10.1007/978-981-19-8555-3_16

Table 1 Satellite communication bands [2]

Frequency	Satellite communication bands	
1575.42 MHz/1227.6 MHz	L-band	GPS
2670–2690 MHz	S-band satellite communication	Transmit
2500–2520 MHz		Receive
4–8 GHz	C-band	
8–12 GHz	X-band	
12–18 GHz	Ku-band	
18–26.5 GHz	K-band	
26–40 GHz	Ka-band	

tracking, forecasting, navigation, surveillance, and astronomy. Satellite communication has lately advanced with the addition of some new services such as high-capacity personal-communication-services (PCS), mobile-communication-satellites (MCS), and direct-broadcast-satellites (DBS) [1]. The most significant component of satellite communication is antenna design. As a result, different satellite services have distinct antenna design requirements for satellite communication applications. To carry multiple applications at the same time, the satellite communication band is further separated into different frequency areas, as shown in Table 1.

C-band, S-band, Ku-band, and Ka-band are the most regularly used band for satellite communication in India. Transponders in these bands are mostly carried by INSAT/GSAT satellites. Traditional satellites operate in medium earth orbit (MEO) (at a height of 900 kms) and have a mass of 500–1000 kg. Furthermore, these satellites are expensive (ranging from 50 to 100 million dollars) and consume a lot of power (eight kilowatts) [3]. Other types of satellites, such as CubeSat that operate in low earth orbits (LEO), lower cost and are more accessible by public. They are less in weight and use only 2 watts of power. They are simpler to construct but have limited utility. Table 2 summarizes the characteristics of both types of satellites and their corresponding examples.

Antenna and repeaters are the main constituents for any satellite service payload as shown in Fig. 1. Antenna is responsible to provide desired radiation characteristics

Table 2 Comparison of conventional (MEO) and cube satellites (LEO) [3]

Type	Conventional	CubeSat
Orbit	MEO	LEO
Mass	50–1000 kg	1–6 kg
Cost	50–100 M	20–200 K
Time to build	4 years	1 year
Power consumption	~800 W	2 W
Examples	Formosat-2	Tokyo Tech 1U (CUTE-1)

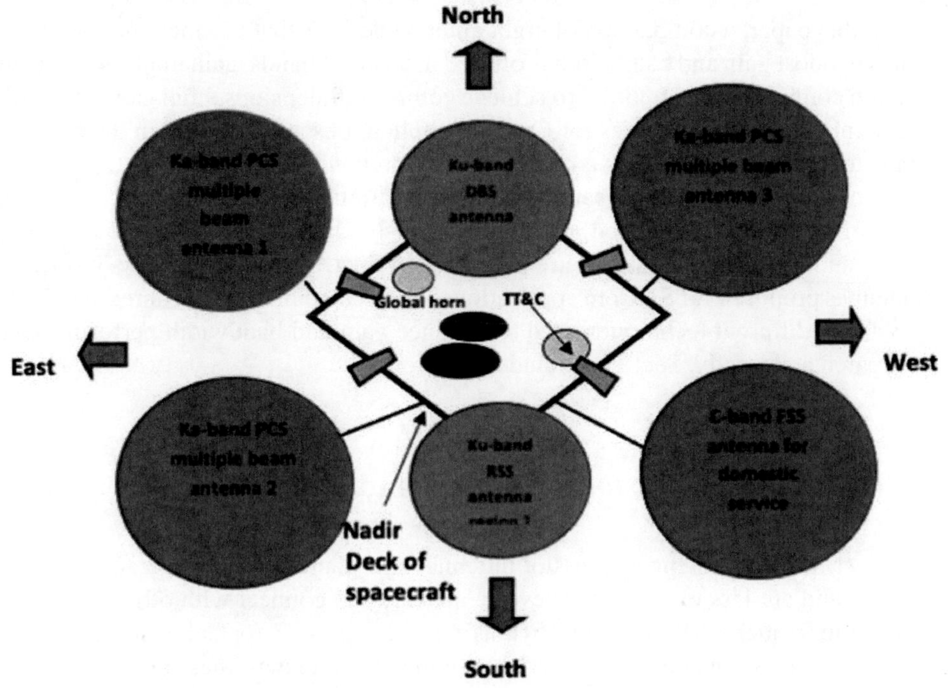

Fig. 1 Multiple satellite services with several supporting antennas [4]

for covering specific region. Thus, they must also exhibit desired polarization and isolation for high cross polar reuse systems.

In addition, antenna gain plays a crucial role in determining the figure of merit for any satellite design which is determined by its gain to temperature (G/T) ratio and effective isotropic radiated power (EIRP) [4]. Different types of antennas used in SatCom applications are summarized in Table 3.

Antennas used for commercial, mobile satellite, and military communication satellites are bulkier and suffer from narrow bandwidth. However, the antenna designs of low earth satellites, i.e. CubeSat, must be smaller owing to light weight of these satellites. The commonly used antennas for CubeSat applications are helical, monopole, slot, and patch antennas.

Table 3 Types of conventional antennas for SatCom applications [4]

Type of antenna	Services
Reflector antennas	Satellite communication. Frequency reuse polarization
Lens antennas and phased antennas	Satellite communication
Phased array antennas	Military communications
Multiple beam antennas	Commercial, mobile, and military satellites

In this paper, a comparison of eight antenna designs that use metamaterial structures to boost gain and bandwidth is offered in terms of bands, gain improvement, and design configurations. In order to achieve compact dimensions, high gain, and wideband, many design strategies for CubeSat applications are proposed in the literature and studied in this work. As a result, this research will contribute to future antenna design trends for traditional satellite communication and CubeSat applications. The remaining paper is organized as follows: The standard configuration of microstrip patch and slot antenna designs are presented in Sect. 2. Further, review of different antennas proposed for SatCom applications is presented in Sect. 3. Thereafter, Sect. 4 discusses different techniques used to enhance gain and bandwidth performance of the patch and finally, Sect. 5 concludes the work.

2 Planar Antennas for Satellite Communication

Microstrip patch antennas and slot antennas are planar in structure and have low profile and are less expensive. They are also easy to connect with other microwave and radio frequency (RF) circuits [5] making them a good fit for CubeSat applications. Furthermore, when compared to reflector and helix antennas, these antennas take up less space and do not require deployment. As a result of these attributes, there is more space available on the CubeSat for solar cells or other equipment. Figures 2 and 3 show the basic patch antenna and slot antenna layouts. The microstrip patch antenna that exhibits resonance in the various satellite communication bands exhibited by Table 1 can be designed using the formulation given in (1)–(5), respectively.

Patch antenna Design Equations [3]

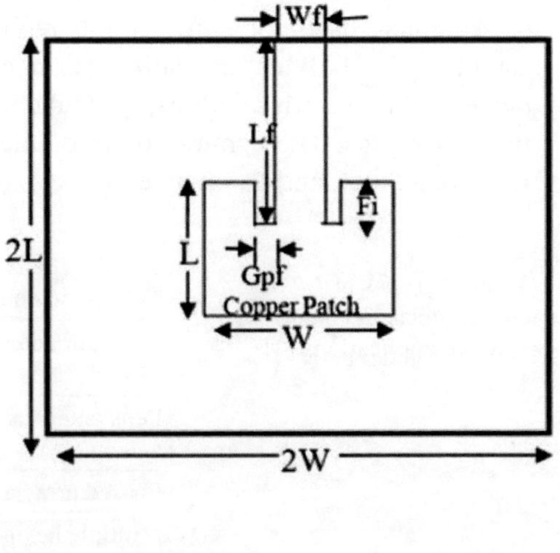

Fig. 2 Microstrip patch antenna [3]

Fig. 3 Slot antenna [3]

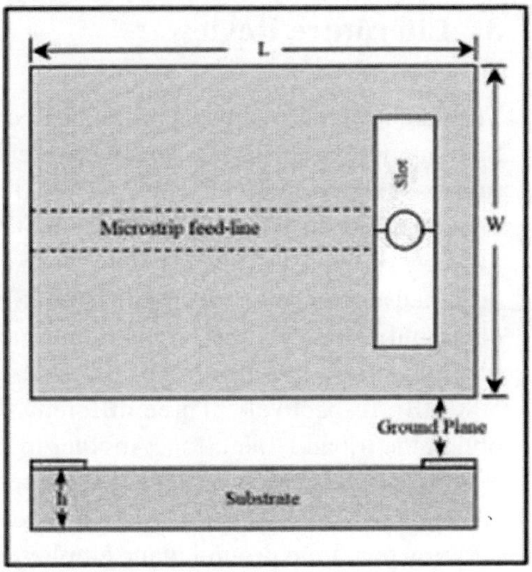

$$W = \frac{C}{2f_r}\sqrt{\frac{2}{\varepsilon_r + 1}} \tag{1}$$

where $\mathcal{E}_r$ is the dielectric constant and c is free space velocity. The $\mathcal{E}_{\text{eff}}$ is the effective dielectric constant of the patch antenna is given by

$$\mathcal{E}_{\text{eff}} = \frac{(\mathcal{E}_r + 1)}{2} + \frac{(\mathcal{E}_r - 1)}{2}\left(1 + 12\frac{h}{W}\right)^{-1/2} \text{where } \frac{w}{h} > 1 \tag{2}$$

The extension length (ΔL) is given as follows:

$$\Delta L = 0.4126\frac{(\mathcal{E}_r + 0.3)}{(\mathcal{E}_r - 0.258)}\frac{\left(\frac{w}{h} + 0.264\right)}{\left(\frac{w}{h} + 0.8\right)} \tag{3}$$

The effective length is further calculated as

$$L_{\text{eff}} = \frac{c}{2f_{r\sqrt{\mathcal{E}_{\text{eff}}}}} \tag{4}$$

Finally, the actual length of the patch is calculated using

$$L = L_{\text{eff}} - 2\Delta L \tag{5}$$

These antennas (patch antenna, slot antenna) may be fed using various feeding techniques such as microstrip feed, coaxial feed, proximity feed, and aperture coupled feed. Each of these feeding techniques has its own advantages and disadvantages.

3 Literature Review

The patch antenna using sapphire dielectric substrate was proposed in [6] for X-band SatCom operation. The S-slots were etched in the patch to exhibit multiband operation resonating at 8.3, 9.2, and 9.8 GHz frequencies. The microstrip feedline structure was employed to achieve impedance matching at the required resonating frequencies. Further, a compact antenna suitable for SatCom applications was designed in [7]. The designed antenna operates for different SatCom applications such as military satellite communication (7.5 GHz) maximum gain of 3.96 dBi, radar applications (9 GHz) having gain of 3.05 dBi, and fifth generation applications (28 GHz) with a gain of 5.86 dBi, respectively. Three different slots were etched on the radiating patch to obtain the triband operation. Another microstrip patch antenna exhibiting dual band operation has been designed in [8] for resonance at 4 GHz and 8 GHz, respectively. Here, two-annular striplines were etched on the top plane of the rectangular-patch in addition to a finite ground plane employed on its bottom layer. The designed antenna exhibited a radiation efficiency of 82.2%, with a measured peak gain of 8.3 and 9.4 dB at 4 and 8 GHz, respectively.

Furthermore, the antenna designed for SatCom applications in [9] resonates at a frequency range of 3.77–13.89 GHz and exhibits ultra-wideband functioning. The wideband operation was achieved by employing Hilbert-shaped metamaterial unit cell array underneath the patch. Also maximum gain of 4.56 dBi and 6.85 dBi was achieved at 5.8 GHz and 8 GHz, respectively, using the designed antenna. Another antenna in [10] comprises of an array of two patch elements for 1U CubeSat applications is shown in Fig. 4.

The detailed review of different patch antennas proposed for CubeSat applications is presented in [3]. Thus, the antenna for satellite communication must exhibit wideband operation [11] and high gain [12] with compact size [13]. Further, to overcome

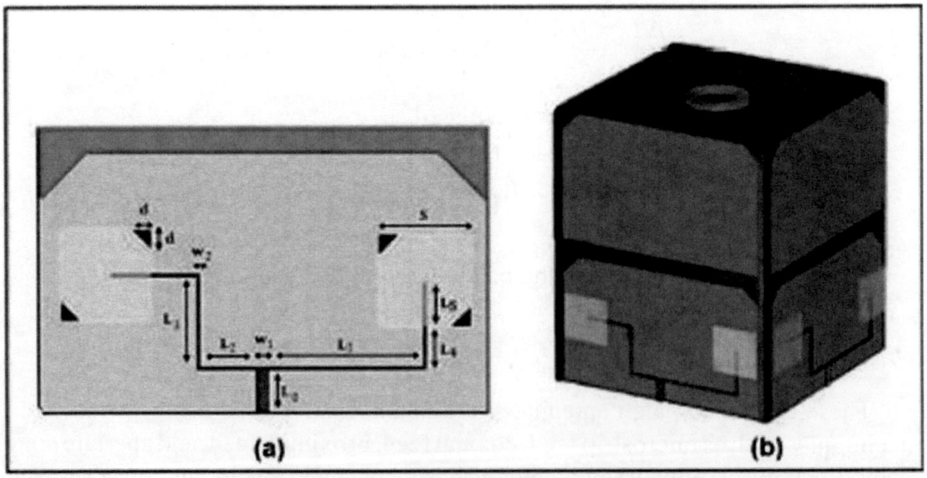

Fig. 4 **a** 2 × 1 patch array antenna, **b** 4 antenna arrays on 1U CubeSat [10]

these challenges, different techniques are proposed for enhancing the gain characteristics and operational bandwidth of the patch antennas for SatCom applications [14–19].

4 Gain and Bandwidth Enhancement Techniques

To achieve high gain multiband operation, patch antennas are loaded with metamaterial substrates or super substrates as in [14]. These materials are extremely helpful in influencing the electromagnetic properties of the antenna parameters. Metamaterials are engineered to have low permittivity/permeability in a specific frequency band, leading in excellent directivity and gain. In addition, the antenna's gain and bandwidth have been improved by employing a multi-split ring resonator metasurface in [15] which acts as a superstrate to enhance its electromagnetic behaviour. Further in [17], the gain of the conventional patch antenna is again enhanced by employing periodic array of single/double-layered cross unit cell metasurface. Further, in [18], remarkable improvement in gain resulted with the addition of three layer superstate metasurface on the traditional patch antenna. Table 4 summarizes various design considerations for attaining high gain multiband antenna design for satellite communication. It is observed by employing metamaterials as substrates, superstates, reasonable improvement in gain, and bandwidth can be achieved.

5 Conclusion

This article provides an overview of the various antenna designs used in satellite communication applications. Patch antennas are loaded with metamaterial structures as a superstrate layer to increase their gain and achieve multiband behaviour. It is demonstrated that metamaterial structures can be used to enhance the patch antenna's electromagnetic characteristics such as gain and bandwidth. As a result, planar patch antennas with the required gain and bandwidth performance proves to be suitable contender for satellite communication applications. By optimizing and tuning the geometrical parameters and shape of metamaterial structure, significant gain improvement can be achieved over varying frequency band.

Table 4 Microstrip patch antennas loaded with metamaterial/metasurface for satellite communication applications

Reference	Type of antenna/technique	Resonant frequencies/bands covered	Gain improvement	Design configuration
[12]	X-band reflect array antenna for satellite communications	8.425 GHz	29.2 dBi	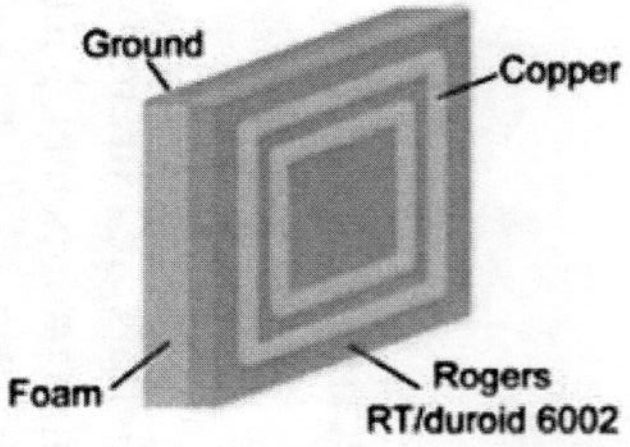
[13]	Dual-polarized corrugated horn antenna	10.2–15 GHz	14–17 dBi gain	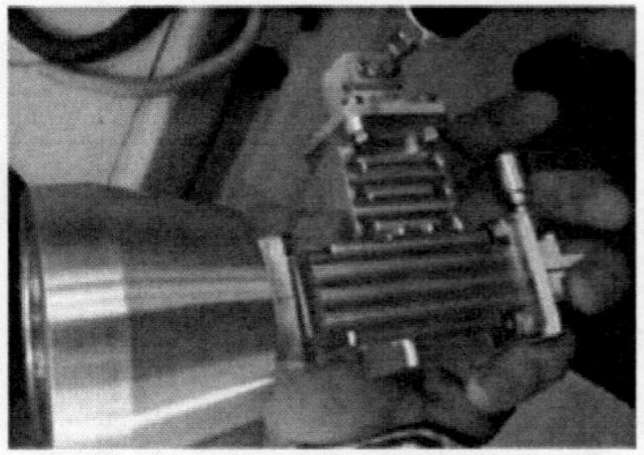

(continued)

Table 4 (continued)

Reference	Type of antenna/technique	Resonant frequencies/bands covered	Gain improvement	Design configuration
[14]	Minkowski fractal-shaped metamaterial (MTM) array	11.6 GHz	2.9 dB	
[15]	Superstrate using multi-split ring resonator (SRR)	5.8 GHz	3.7 dBi	

(continued)

Table 4 (continued)

Reference	Type of antenna/technique	Resonant frequencies/bands covered	Gain improvement	Design configuration
[16]	Metamaterial as a single- or a double-layer superstrate	43 GHz	5.1 dB (single layer), 7 dB (double layer)	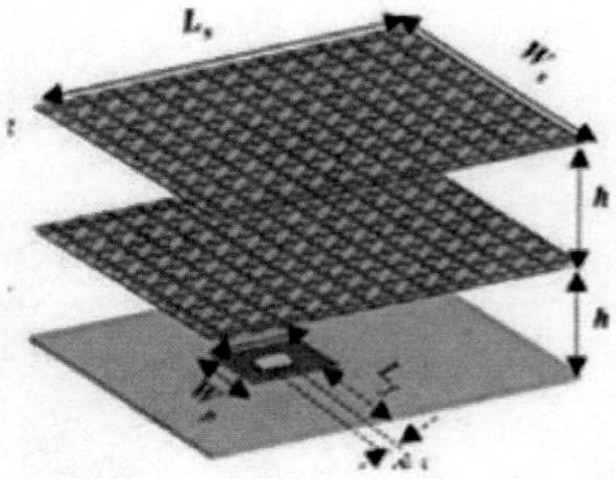
[17]	Three layer superstrate	5.9 GHz	2.69 dB	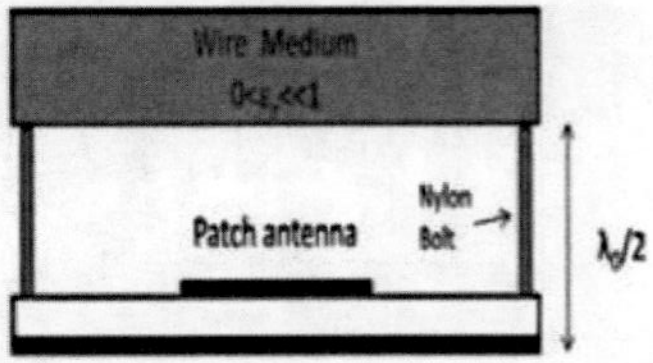
[18]	Double-inverse-epsilon-shaped, triple-band epsilon-negative (ENG) metamaterial	S-band (2.38 GHz), C-band (4.55 GHz), X-band (9.42 GHz)	–	

(continued)

Table 4 (continued)

Reference	Type of antenna/technique	Resonant frequencies/bands covered	Gain improvement	Design configuration
[19]	Dual-band double negative metamaterial structure employing flexible nickel aluminate ($NiAl_2O_4$)	X-band (8.34 GHz) Ku-band (12.78 GHz and 14.32 GHz)	–	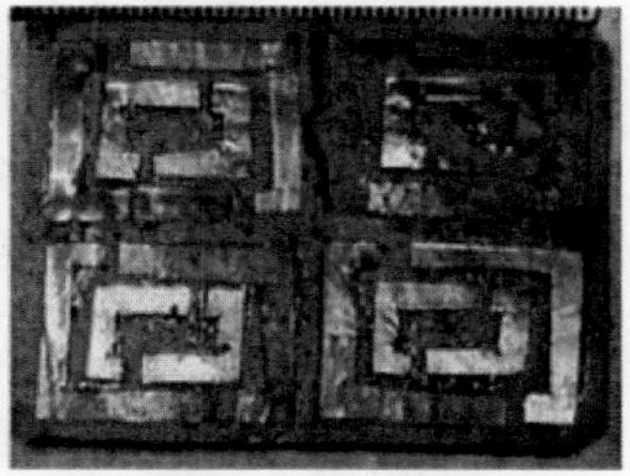

References

1. Prasad PC, Chattoraj N (2013) Design of compact Ku band microstrip antenna for satellite communication. In: 2013 International conference on communication and signal processing. IEEE, pp 196–200
2. Rao KC, Rao PM, Naidu ML, Adiarayana V (2022) A compact rectangular dual patch antenna for multiple satellite communication applications. Wireless Pers Commun 122(2):1007–1041
3. Antenna Designs for CubeSats: A Review4
4. Sharma M (2020) Design and analysis of MIMO antenna with high isolation and dual notched band characteristics for wireless applications. Wireless pers commun 112:1587–1599. https://doi.org/10.1007/s11277-020-07117-4
5. Kumar G, Singh D, Kumar R (2021) A planar CPW fed UWB antenna with dual rectangular notch band characteristics incorporating U-slot, SRRs, and EBGs Int J RF Microw Compute Aided Eng 31(7)
6. Nishandhi S, Reddy DY, Ajith Kumar R (2022) Designing of microstrip patch antenna for satellite communication. In: Futuristic communication and network technologies. Springer, Singapore, pp 167–178
7. El Hadri D, Zugari A, Zakriti A, El Ouahabi M, Taouzari M (2020) A compact triple band antenna for military satellite communication, radar and fifth generation applications. Adv Electromagnet 9(3):66–73
8. Hassan MM, Hussain M, Khan AA, Rashid I, Bhatti FA (2021) Dual-band B-shaped antenna array for satellite applications. Int J Microw Wirel Technol 13(8):851–858
9. Elwil TA, Jassim DA, Mohammed HH (2020) Novel miniaturized folded UWB micro strip antenna-based metamaterial for RF energy harvesting. Int J Commun Syst 1(2)
10. Maged MA, Elhefnawi F, Akah HM, El-Hennawy HM (2018) C-band transparent antenna design for intersatellites communication. Int J Sci Eng Res 9(3):248–252
11. Koli MNY, Afzal MU, Esselle KP, Hashmi RM (2020) An all-metal high-gain radial-line slot-array antenna for low-cost satellite communication systems. IEEE Access 8:139422–139432. https://doi.org/10.1109/ACCESS.2020.3012787
12. Ma B, Lu F, Zhi G, Xue X, Zhao X, Ma C, Fan Y, Yang M (2021) Development of an X-band reflectarray antenna for satellite communications. Sci Rep 11(1):1–9
13. Manshari S, Koziel S, Leifsson L (2020) Compact dual-polarized corrugated horn antenna for satellite communications. IEEE Trans Antennas Propag 68(7):5122–5129. https://doi.org/10.1109/TAP.2020.2980337
14. Gupta N, Saxena J, Bhatia KS, Dadwal N (2019) Design of metamaterial-loaded rectangular patch antenna for satellite communication applications. Iranian J Sci Technol Trans Electr Eng 43(1):39–49
15. Arora C, Pattnaik SS, Baral RN (2016) Metamaterial superstrate for performance enhancement of microstrip patch antenna array. In: 2016 3rd International conference on signal processing and integrated networks (SPIN), February. IEEE, pp 775–779
16. Bouzouad M, Chaker SM, Bensafielddine D, Laamari EM (2015) Gain enhancement with near-zero-index metamaterial superstrate. Appl Phys A 121(3):1075–1080
17. Gangwar D et al (2011) Enhancement of front to back ratio and directivity with wire medium Epsilon-Near zero metamaterial as superstrate in microstrip patch radiators. In: IEEE Indian antenna week—workshop on advanced antenna technology. IEEE, Kolkata, India, pp 6–9. https://doi.org/10.1109/indianaw.2011.6264936
18. Afsar MSU, Faruque MRI, Khandaker MU, Alqahtani A, Bradley DA (2022) A new compact split ring resonator based double inverse epsilon shaped metamaterial for triple band satellite and radar communication. Curr Comput-Aided Drug Des 12(4):520
19. Faruque MRI, Ahamed E, Rahman MA, Islam MT (2019) Flexible nickel aluminate ($NiAl_2O_4$) based dual-band double negative metamaterial for microwave applications. Results Phys 14:102524

Conversion Efficiency Enhancement of Amorphous-Si:H Solar Cell for Space Satellite Antenna Applications

Shivani Malhotra, Lipika Gupta, Jaya Madan, and Hritik Nandan

Abstract Hybrid combination of Amorphous-Si (a-Si) solar cells with antennas (Solants) has important use in space satellites to solve the dual purpose of transmitting and receiving electromagnetic waves while generating power. This paper describes a stacked a-Si-based solar cell which can be easily integrated with a stacked microstrip antenna thus enhancing its technological capability. The intrinsic layer (*i*-layer) thickness of a *p-i-n* solar cell is varied from 0.1 and 1 μm for space and power optimization. The optimal thickness for maximum power conversion efficiency is obtained to be 0.3 μm. After the solar cell is integrated with the antenna, it does not alter the overall structure, thereby, a lot of space is saved through the innovative technology design. This integration also provides an enhance power generation capability for dependable and enduring satellite communication.

Keywords Antenna · *p-i-n* solar cell · Solant · Amorphous-Si:H solar cell · Stacked structure

1 Introduction

The majority of developed countries as well as countries still developing their economies have shifted their focus to the investigation of alternative and renewable forms of energy. Solar energy utilization is at the forefront of these efforts. This is because of its abundance, inexhaustibility, and pollution-free nature. The scientific discovery of the photoelectric effect has solved many energy problems. We have no need to search further afield for a source of energy so long as the sun continues to illuminate the sky. Converting science into technology is the issue that must be addressed. Solar photovoltaic is one of the prominent sources of clean and green

S. Malhotra (✉) · L. Gupta · H. Nandan
Chitkara University Institute of Engineering & Technology, Chitkara University, Rajpura, Punjab 140401, India
e-mail: shivani.malhotra@chitkara.edu.in

J. Madan
VLSI Centre of Excellence, Chitkara University Institute of Engineering and Technology, Chitkara University, Rajpura, Punjab, India

D. Singh et al. (eds.), *Computer Aided Constellation Management and Communication Satellites*, Lecture Notes in Electrical Engineering 987,
https://doi.org/10.1007/978-981-19-8555-3_17

energy. Recently, research has concentrated on developing higher conversion efficiency solar cells with lower manufacturing cost. Different material properties, material combinations, design and procedure of fabrication are some of the factors which influence the photoactivity in solar cells. The researchers investigated the integration of solar photovoltaic (PV) systems with a variety of other systems, including power grids, wearable technology, building structures such as windows and roof tops, space satellites, and solant, among others. A solant, which is a hybrid of a solar cell and an antenna, has garnered considerable attention due to its dual function of transmitting and receiving electromagnetic waves while also generating power. Significant effort has been made to improve the performance of solants for small satellites [1, 2]. The primary design challenge is to integrate the solar cell with the microstrip patch, or slot antennas in such a way that the antenna's performance remains constant while the solar cell's efficiency is maximized [3]. Although the fabricated solar cells achieved close to 30% efficiency, the integration of solar cells with antennas is an intriguing problem in terms of obtaining the best energy conversion efficiency with the change in the width and thickness of the solar cell and interference with antenna parameters [4].

Vaccaro et al. [5] and Zhang et al. [6] described a solar cell built on amorphous silicon (a-Si), with a printed slot antenna. Based on thin-film-technology, amorphous-Si solar cells has low processing cost [7], and are easily adapted with the shape of the antenna and can be easily grown over the antenna surface. Podilchak et al. [8] considered placing a square patch antenna on satellites fully integrated with a solar cell layer and transparent glass substrate. On the similar lines, Jones et al. [9] demonstrated through the findings that mounting solar cell atop an aperture coupled microstrip antenna has no noticeable influence on antenna operations and at the same time it provided significant space efficiency. Ali et al. [10] developed three-layer solar cells with integrated antennas for wireless applications using copper indium gallium selenide (CIGS). Using an alternative approach, Tawk et al. [11] vertically mounted the copper-based F-antenna array atop with the solar cells without impacting its performance. Thus, the literature focuses on antenna-solar cell integration in terms of space optimization and antenna parameter interference. However, optimization of the solar cell structure is necessary to achieve the highest conversion efficiency possible. Thus, the literature is concentrated on antenna-solar cell integration in terms of space optimization and interference with antenna parameters. However, it is necessary to optimize the solar cell structure for achieving highest possible conversion efficiency.

Through this article, an efficient solar cell structure for space applications has been proposed. The motivation is to investigate the outcome of varying the width of intrinsic layer (*i*-layer) on the different parameters of amorphous-Si:H-based PIN solar cell. Further, the solar cell with the optimal thickness aids in space optimization in solant. Major parameters under consideration are power conversion efficiency (PCE), open-circuit-voltage (V_{oc}) and current–density (J_{sc}). Section 2 of the article discusses the structure of solar cell devices. Section 3 discusses the findings, while Sect. 4 draws the conclusion.

2 Device Structure and Methodology

The current study is conducted using the simulation software Solar Cell Capacitance Simulator (SCAPS). SCAPS is being developed at the University of Ghent's Department of Electronics and Information Systems (ELIS) [12]. The simulator provides a best approach to model and optimize the solar cell parameters. The a-Si solar cell is used in this antenna due to its mature technology and being cost-effective in processing. Additionally, the integrated solar cell design does not alter the antenna's inherent structure and maximize power/weight ratio. The amorphous-Si:H (hydrogenated amorphous-silicon) cell is stacked in a manner similar to that of a microstrip slot antenna. The cells are flexible films that can easily be shaped into a patch or an array. These cells can be easily fabricated on different materials. Hence, solar cells and microstrip slot antennas are combined.

Figure 1 [6] illustrates the device structure of a simulated solar cell FTO/*p* > amorphous-Si:H (0.009 μm)/*i* > amorphous-Si:H (0.500 μm)/*n* > amorphous-Si:H (0.020 μm)/Au. The important solar cell layers are formed with 0.009 μm highly conductive *p*-layer, followed by undoped 0.500 μm thick intrinsic layer responsible for maximum absorption of photons and 0.020 μm thick highly conductive *n*-layer. A PIN-type structure is developed to facilitate charge carrier separation and to improve the antenna's overall performance. Additionally, adjustable energy bandgap of amorphous-Si:H helps to attain higher energy conversion efficiency due to high absorption coefficient. Typical band gap of amorphous-Si:H varies from 1.65 to 1.88 eV [13]. As a result of this motivation, the intrinsic layer thickness is varied between 0.1 and 1 μm to achieve the optimal design parameters for the highest performing solar cell.

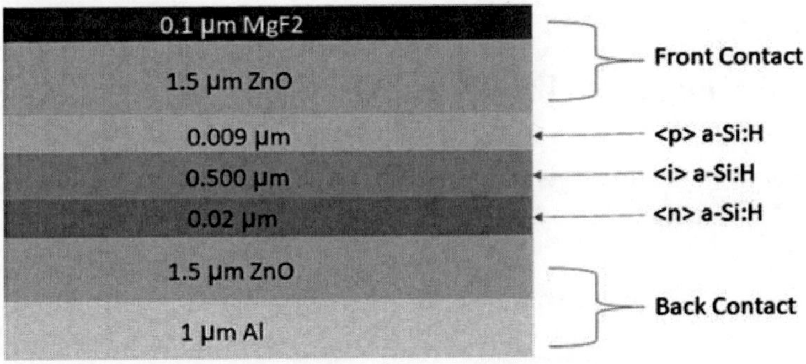

Fig. 1 Device structure of amorphous-Si:H solar cell simulated on SCAPS 1D in this work

3 Results and Discussion

3.1 Effect of Intrinsic Layer Thickness on Photovoltaic Parameters

The influence of change in width of intrinsic layer in the range of 0.1–1 μm on photovoltaic parameters using p and n-type materials is depicted in Fig. 2. These findings demonstrate that the PV parameters are depend on the intrinsic layer width. Optimizing the *i*-layer thickness not only critically improves the solar cell's performance but also improves overall design of the antenna. As illustrated in Fig. 2, the PCE decreases exponentially as the thickness exceeds 0.4 μm. Similar depreciation behavior can be observed in V_{oc} and Fill-Factor (FF) when the intrinsic layer thickness is increased beyond 0.4 μm. On the other hand, changes in thickness (0.4–1 μm) had a negligible effect on the J_{sc}, rendering it resistant to thickness changes, while a slight dip was observed after 1 μm. Following a thorough examination of Fig. 2, it is determined that the optimal thickness for maximum efficiency is 0.3 μm. PV parameters all perform admirably at the same thickness. The PV parameters thus attained for the simulated solar cell structure are: PCE = 12.04%, $J_{sc} = 17.5$ mA/cm^2, $V_{oc} = 1.02$ V, FF = 82.5%. The obtained parameters are comparable with the similar amorphous-Si:H PIN solar cell reported by Bechane et al. in 2020 [14]. The proposed solar cell parameters also compared with the [15] wherein the solar cell structure w.r.t. *p-i-n* layer thickness is same. It is pragmatic that the PV parameters are in approximation except FF which is higher if the optimal thickness of *i*-layer is considered as 0.3 μm. Table 1 gives the comparison of the PV parameters obtained for the solar cell considered in this work with solar cells considered in [14, 15].

3.2 Effect on Power Voltage Curve

The power voltage curve also showed the best results when width of the intrinsic layer is 0.3 μm. The performance in terms of power and voltage can be seen in Fig. 3a, b.

4 Conclusion

In this paper, amorphous-Si:H PIN solar cell stacked structure was considered which can be used with the stacked microstrip antenna (solant) in space satellites. The width of the i-layer of amorphous-Si:H is varied from 0.1 to 1.0 μm to get the optimum power conversion efficiency and space optimization. The stacked geometry of amorphous-Si:H solar cell can be perfectly combined with an antenna without

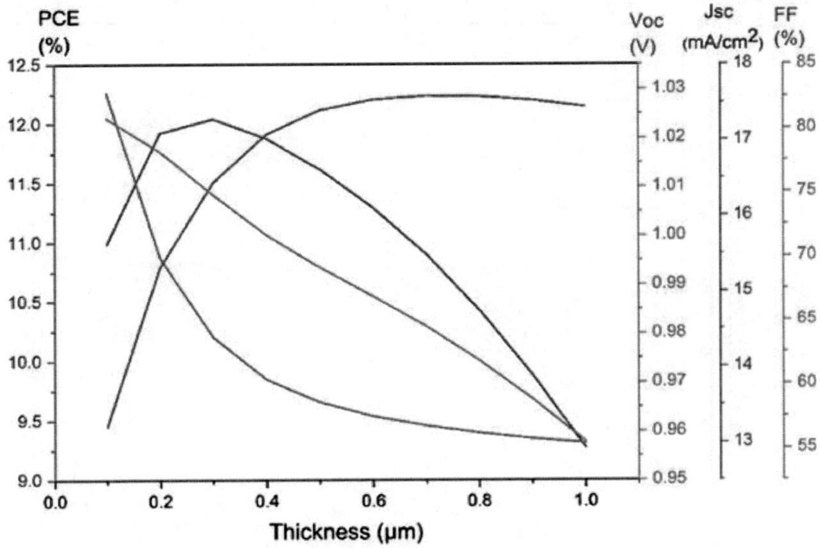

Fig. 2 Effect of intrinsic thickness on PV parameters

Table 1 Comparison of the PV parameters

Parameter	[14]	[15]	This work
Device Structure	TCO/*p* > a-Si:H (0.01 μm)/*i* > a-Si:H (0.3–0.6 μm)/*n* > a-Si:H (0.01 μm)/metal	TCO/*p* > a-Si:H (0.009 μm)/*i* > a-Si:H (0.5 μm)/*n* > a-Si:H (0.02 μm)/metal	FTO/*p* > a-Si:H (0.009 μm)/*i* > a-Si:H (0.1–1 μm)/*n* > a-Si:H (0.02 μm)/Au
i-layer thickness variation	0.3–0.6 μm	Fixed	0.1–1 μm
PCE	12.655%	11.59%	12.04%
J_{sc}	13.145 mA/cm^2	17.36 mA/cm^2	17.5 mA/cm^2
V_{oc}	1.193 V	0.965 V	1.03 V
FF	80.7%	69.1%	82.5%
Optimum thickness of *i*-layer	0.6 μm	0.5 μm	0.3 μm

affecting the antenna parameters. The integration further enhances the power generation capability for an optimum thickness of *i*-layer as 0.3 μm. The simulated PV parameters attained for the solar cell structure are: The simulation revealed the obtained parameters PCE = 12%, J_{sc} = 17.5 mA/cm^2, V_{oc} = 1.02 V, FF = 82.5% are in coherence with standard amorphous-Si:H solar cell. The stacked solar cell structure considered in this work, on the other hand, is suitable for integration with antennas to be used in space application.

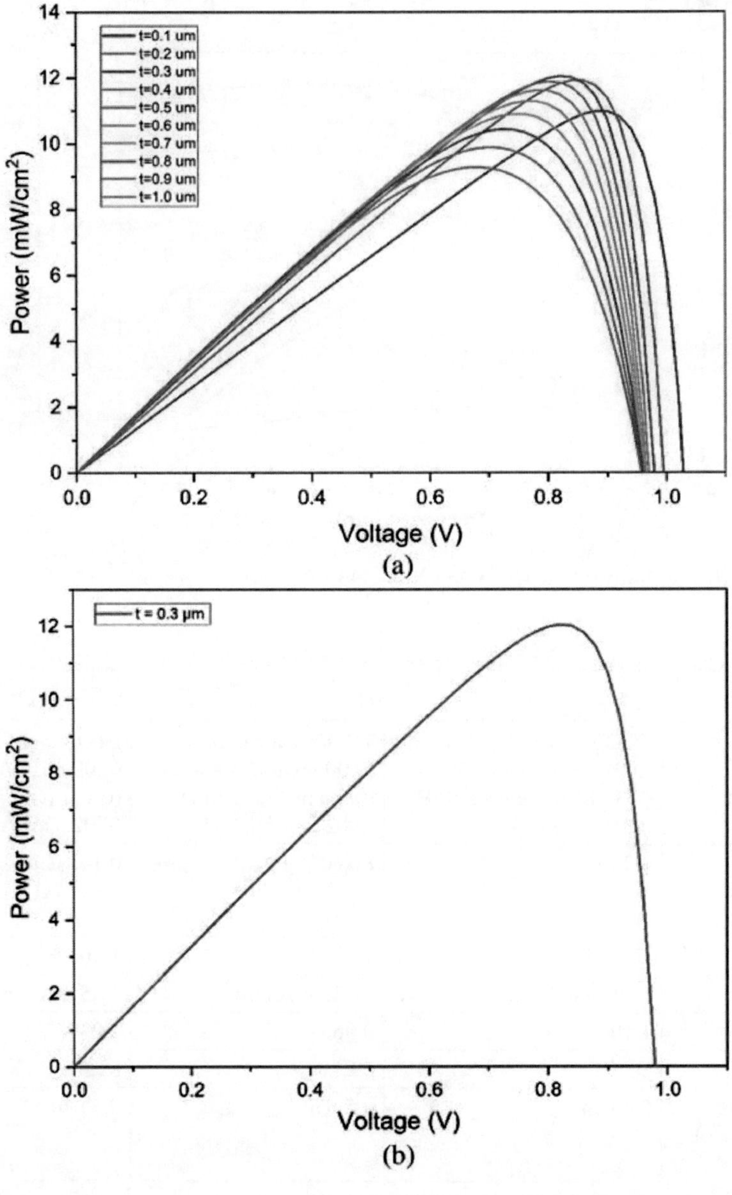

Fig. 3 Power voltage curve **a** for the thickness of intrinsic layer from 0.1 to 1.0 μm, **b** for best thickness $t = 0.3$ μm

References

1. Thandullu Naganathan SB, Dhandapani S (2022) Patch antenna integrated on solar cells for green wireless communication: a feature oriented survey and design issues. Int J RF Microw Comput Eng 32(1). https://doi.org/10.1002/MMCE.22926
2. Kong M et al (2020) Survey of energy-autonomous solar cell receivers for satellite–air–ground–ocean optical wireless communication. Prog Quantum Electron 74. https://doi.org/10.1016/J.PQUANTELEC.2020.100300
3. Zhao Y, An W, Luo Y, Li SR, Xiong L, Yu S (2021) Low-profile antenna integrated with solar cells for the 2.4 GHz band. IEEE Antennas Wirel Propag Lett 20(4):443–447. https://doi.org/10.1109/LAWP.2021.3051795
4. Naresh B, Singh VK, Sharma VK (2021) Integration of RF rectenna with thin film solar cell to power wearable electronics. Int J Microw Wirel Technol 13(1):46–57. https://doi.org/10.1017/S1759078720000410
5. Vaccaro S, Mosig JR, De Maagt P (2003) Two advanced solar antenna 'SOLANT' designs for satellite and terrestrial communications. IEEE Trans Antennas Propag 51(8):2028–2034. https://doi.org/10.1109/TAP.2003.815424
6. Zhang Z, Bai B, Li X, Liu Y, Sun C, Zhang Y (2020) Integration of circularly polarized microstrip slot array antenna with amorphous silicon solar cells. IEEE Antennas Wirel Propag Lett 19(12):2320–2323. https://doi.org/10.1109/LAWP.2020.3031608
7. Singh S, Kumar S, Dwivedi N (2012) Band gap optimization of p–i–n layers of a-Si: H by computer aided simulation for development of efficient solar cell. Sol Energy 86(5):1470–1476. https://doi.org/10.1016/J.SOLENER.2012.02.007
8. Podilchak SK, Comite D, Montgomery BK, Li Y, Gomez-Guillamon Buendia V, Antar YMM (2019) Solar-panel integrated circularly polarized meshed patch for Cubesats and other small satellites. IEEE Access 7:96560–96566. https://doi.org/10.1109/ACCESS.2019.2928993
9. Jones TR, Grey JP, Daneshmand M (2018) Solar panel integrated circular polarized aperture-coupled patch antenna for cubesat applications. IEEE Antennas Wirel Propag Lett 17(10):1895–1899. https://doi.org/10.1109/LAWP.2018.2869321
10. Ali A, Wang H, Lee J, Ahn YH, Park I (2021) Ultra-low profile solar-cell-integrated antenna with a high form factor. Sci. Reports 11(1):1–9. https://doi.org/10.1038/s41598-021-00461-w
11. Tawk Y, Costantine J, Ayoub F, Christodoulou CG (2018) A communicating antenna array with a dual-energy harvesting functionality [wireless corner]. IEEE Antennas Propag Mag 60(2):132–144. https://doi.org/10.1109/MAP.2018.2796025
12. Burgelman M, Nollet P, Degrave S (2000) Modelling polycrystalline semiconductor solar cells. Thin Solid Films 527–532. Available: www.elsevier.com/locate/tsf. Accessed 23 Apr 2022 (Online)
13. Kabir MI, Shahahmadi SA, Lim V, Zaidi S, Sopian K, Amin N (2012) Amorphous silicon single-junction thin-film solar cell exceeding 10 % efficiency by design optimization. Int J Photoenergy 2012. https://doi.org/10.1155/2012/460919
14. Bechane L, Bouarissa N, Loucif K (2020) Numerical simulation and optimization of the performances of a solar cell (p-i-n) containing amorphous silicon using AMPS-1D. Trans Electr Electron Mater 22(4):531–535. https://doi.org/10.1007/S42341-020-00262-4
15. Ayat L, Nour S, Meftah A (2019) Comparative study of a PIN homojunction a-Si: H solar cell. J. Ovonic Res. 15(1):89–94

Analytical Review on Satellite Communication: Benefits, Issues, and Future Challenges

Nishant Tripathi, Kamal Kumar Sharma, and Utkarsh Pandey

Abstract Since 1945, when Arther C. Clarke proposed the synchronous motion of satellites with constant angular velocity in lockstep with the earth's rotation, satellite communication has been a reality. Satellite communications have recently experienced a renaissance in popularity, owing to advancements in technology and private investment. These findings will aid in the direction of future satellite communication research. Onboard processing and data collection are just a few of the primary drivers of innovation while earth observation, and aviation and maritime tracking and communication are few add on area of research applications. The following five axes are discussed in detail: the origins of satellite communication, the components of a satellite link, the advantages and disadvantages of satellite communication, frequency allocation, and the current situation.

Keywords LEO · GEO · Bandwidth · Satellite communication

1 Introduction

According to the elemental approach of the satellite communication block diagram, it may consist of numerous ground stations that are connected to a space-based satellite [1–3]. A terrestrial network connects the user to the earth station [1–3]. To avoid interference, the frequency spectrum for the downlink should be distinct from the frequency spectrum for the uplink. The signal is processed at the receiving earth station to obtain the base band signal, which is then transmitted to the user via a terrestrial network.

N. Tripathi (✉) · K. K. Sharma · U. Pandey
Department of Electronics and Communication Engineering, Lovely Professional University, Phagwara, Punjab, India
e-mail: phd.nishant17@gmail.com

K. K. Sharma
e-mail: kamal.23342@lpu.co.in

N. Tripathi · U. Pandey
Pranveer Singh Institute of Technology, Kanpur, India

D. Singh et al. (eds.), *Computer Aided Constellation Management and Communication Satellites*, Lecture Notes in Electrical Engineering 987,
https://doi.org/10.1007/978-981-19-8555-3_18

Additionally, modern communication satellites make use of frequency reuse to maximize the number of transponders within the available bandwidth [1]. The 6/4 GHz band is the most popular due to its low propagation difficulty. At 4 GHz, rain attenuation and sky noise are low, making it possible to construct a receiving system [1] (Figs. 1, 2, 3 and 4).

The signal is processed through an LNA and then down converted, demodulated, and decided by a decoder, yielding the original baseband signal. In the event that uninterruptible power cannot be maintained, some critical components will be installed

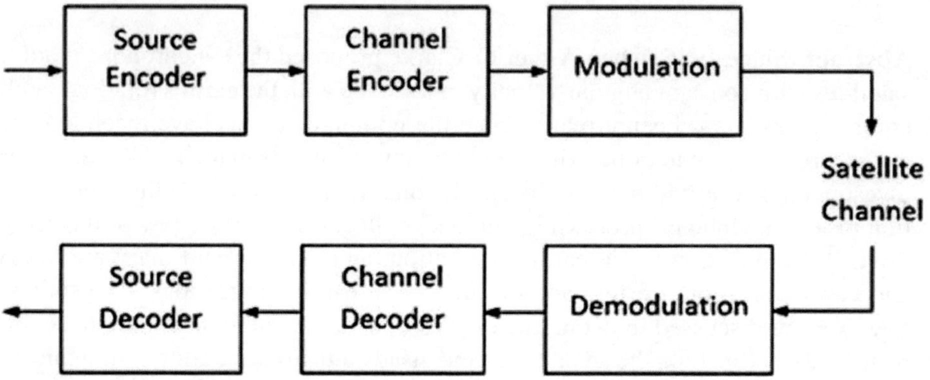

Fig. 1 Satellite communication model

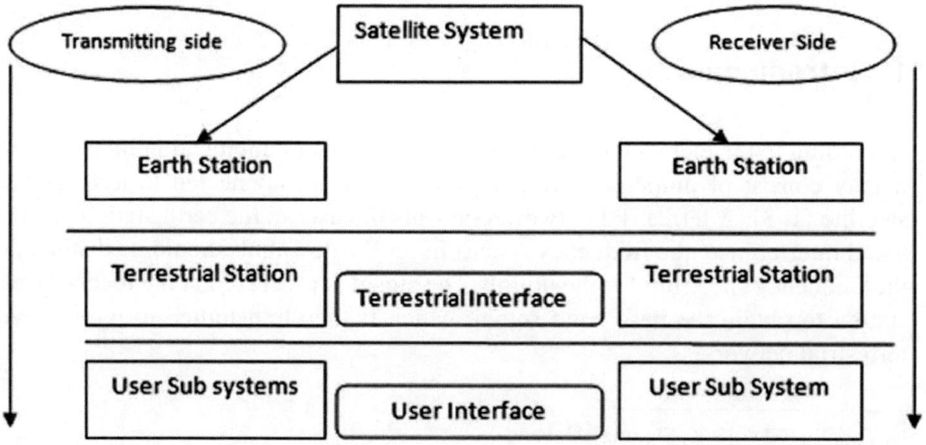

Fig. 2 Elements of a satellite communication link

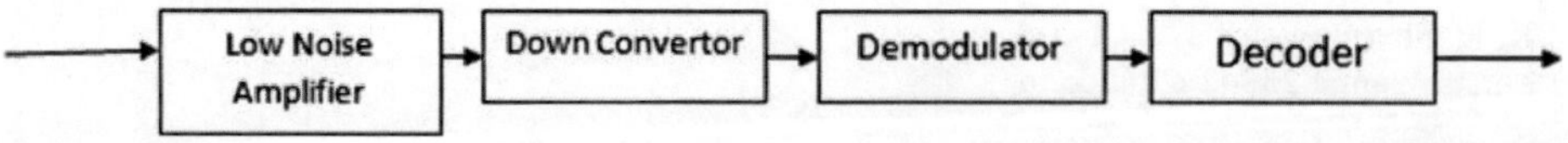

Fig. 3 Transmitter [4] of earth station

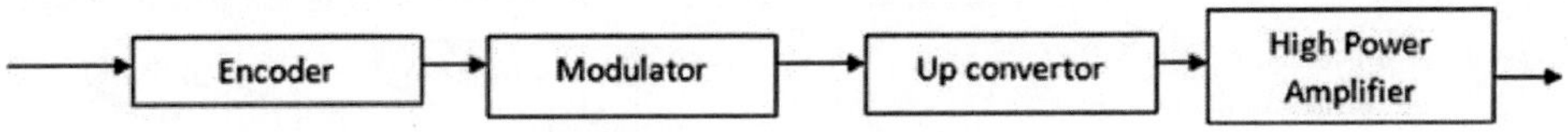

Fig. 4 Receiver [4] of earth station

and an automatic switchover will occur. The isolation of HPA and LNA is critical when designing an earth station.

1.1 Advantage and Disadvantage of 6/4 GHz Band

- No rain absorption
- Low propagation problem
- No polarization changes during wave passing through ionosphere low sky noise makes SNR low at receiving antenna.
- Broad beam width allowed large coverage area at receiver
- Interference is higher from other users power wastage is high
- Direct reception at home is tough

2 Satellite Frequency Allocation and Band Spectrum

Satellite frequency allocation is very difficult job in hand. This is done by International Telecommunication Union (ITU) making three sub-divisions to allocate frequency region wise:

- Coverage1:- Europe, Africa, what was formerly the Soviet Union and
- Mongolia. [1]
- Coverage 2:-North America, South America, and Greenland [1] Coverage 3:- Asia, Australia, and South West Pacific [1]

  Some of the basic services provides by the satellite are:-

- Fixed satellite services [5, 6]
- Broadcasting satellite services [5, 6]
- Mobile satellite services [5, 6]
- Navigational satellite services [5]
- Meteorological satellite services [5]
- Biomedical satellite services [5]
- Topographical satellite services [5]
- Military operation satellite services [5] (Tables 1, 2 and 3)

Table 1 Frequency bands for satellite communication [7]

Band	Downlink band (MHz)	Uplink band (MHz)
UHF-military	Between 250–280	Between 290–310
C band-commercial	Between 3700–4300	Between 5900–6450
X band-military	Between 7000–7800	Between 7900–8400
Ku band-commercial	Between 11,700–12,200	Between 14,000–14,600
Ka band-commercial	Between 17,000–21,200	Between 27,000–30,000
Ka band-military	Between 20,200–21,200	Between 43,500–46,000

Table 2 Frequency band designation

Frequency (MHz)	General mobile
806–890	Region 2 and 3
1530–1535	Limited use
1544–1545	Down
1545–1559	Down
1645–1646	UP
29.5–31.0	UP
39.5–40.5	Down
43.5–47	UP
66–71	UP/down
71–74	UP/down
81–84	UP/down
95–100	UP/down
134–142	UP/down
190–200	UP/down
250–265	UP/down

3 Advantage and Disadvantage of Satellite Communication

3.1 Advantages

- Communication between multiple points is possible.
- Satellite circuits can be installed quickly.
- During failures or critical conditions, each earth station can be quickly removed from its current location and reinstalled elsewhere.
- Flexible communication enables seamless mobile communication.

Table 3 Frequency allocation for MSS

Frequency range (GHz)	Band designation [8]
0.1–03	VHF [8]
0.3–1.0	UHF [8]
1.0–2.0	L [8]
2.0–4.0	S [8]
4.0–8.0	C [8]
8.0–12.0	X [8]
12.0–18.0	Ku [8]
18.0–27.0	K [8]
27.0–40.0	Ka [8]
40.0–75.0	V [8]
75.0–110.0	W [8]
110.0–300.0	Mm [8]
300–3000	μm [8]
0.1–03	VHF [8]
0.3–1.0	UHF [8]

- Because the cost of a satellite is independent of distance, it is cost effective.
- Superior signal transmission combined with increased privacy and security.

3.2 Disadvantages

- Between signal transmission and reception, there is a fractional second delay. Efficiency is insufficient.
- Any file transfer takes longer.
- Low antenna gain results in bandwidth overcrowding. Atmospheric losses are high above 30 GHz.
- Satellites of various types satellite of the natural world

4 Application-Based Satellite

Astronomical Used to observe distant planets, stars, galaxies, and other objects in the universe. It is a space telescope suspended in orbit for the purpose of photographing objects in space.

Biosatellite Involves the placement of animals or plants in space in order to conduct research on the effects of space on these living things.

Telecommunications These satellites provide telecommunications services. Telecasting, telephone calls, Internet connectivity, radio broadcasting, and a great deal of remote connectivity are all examples of typical applications.

Earth Observation Deployed for non-military purposes to study the environment, monitor climatic changes, and map the earth.

Navigation Enables the precise location of any object on the earth to be determined. This results in the creation of novel applications, technologies, and business models. Miniaturized at a low cost, smaller, and lighter satellites are launched for the limited purpose of scientific data collection and radio relay.

5 Current Research Initiatives, Future Directions, and New Developments in Satellite Communication (2019–2022)

Yi Lung Mo [5, 9] of the University of Houston in the United States of America presented a paper titled Periodic material-based vibration isolation for satellites [5, 9]. Applications of satellite remote sensing for hydrologic studies were presented by Hyongki Lee, a student at the University of Houston [5, 9]. The title of this paper is "A time domain along-Track SAR interferometer method" by Ning Cao of the University of Houston in the United States of America [5, 9]. Boris A. Khrenov Lomonosov is a composer from Russia. Moscow State University is a public research university located in the Russian capital of Moscow, Russia. This study examines the phenomenology of near UV flashes in the earth's atmosphere as observed by satellite, as well as their relationship to thunderstorm regions [9].

The Dorian Gorgan Technical University in Cluj-Napoca, Romania is a private institution located in Europe. The title of this paper is Processing Earth Observation Data in a Flexible and Adaptive Manner Using High-Performance Computation Architectures [2, 9, 31, 32]. Lt. Abdullah Kaya is a military officer at the Turkish Air War College (Turkish Air War College) in Turkey (Ali Emre Destegul, College of Turkish Air Warfare). This article discusses the use of satellites in border security in Turkey [2, 9, 10, 11, 12]. The Yoshinari Minami Advanced Science-Technology Research Organization is a 501(c) (3) not-for-profit organization dedicated to advanced science-technology research.

Vladislav V Demyanov is the name of Irkutsk State Transport University. The title of this paper is Russia's GNSS positioning availability control in the face of space weather hazards. National Aeronautics and Space Administration (NASA), William Walker, United States of America Lithium-ion batteries are being tested and simulated in radiation-driven space environments using thermoelectric and electrochemical methods.

Robert P. Whearty Marsh Space Projects in the United States This article discusses satellite launch and in-orbit insurance. Ali Cheknane is a Tunisian writer and musician. Algeria's Amar Telidji University is a prestigious institution. The title describes a method for defect detection and yield control of grid-connected photovoltaic systems using satellite monitoring. Aerospace engineers [13, 14] have developed a method for using articulated solar panels to steer the satellite during aero braking, reducing the number of passes required and potentially saving propellant, time, and money.

6 Conclusion

Satellite communications [1] have recently entered a critical stage of their evolution, owing to the explosive growth of various Internet-based applications and services, which has resulted in an ever-increasing demand for broadband high-speed, heterogeneous, ultra-reliable, and low-latency communications systems and networks [1]. Satellites, with their unique characteristics and technological advancements, can contribute significantly to meeting this demand, either as a stand-alone solution or as part of a larger satellite-terrestrial network integration effort [1]. The authors of this paper have compiled the most recent technological advances in scientific, industrial, and standardization analyses in the field of satellite communications in order to accomplish several review aspects for the future researchers [1, 5].

References

1. Huang J, Cao J, Liang Q, Wang W, Mu J, Liu X, Na Z (2020) Recent development of commercial satellite communications systems. Lect Notes Electr Eng 572
2. Livieratos SN, Ginis G, Cottis PG (1999) Availability and performance of satellite links suffering from interference by an adjacent satellite and rain fades. IEE Proc Commun 146(1):61–67
3. Olsen RL (1993) Interference due to hydrometeor scatter on satellite communication links. In: ITU-R, RP (ed) Proc IEEE, vol 81, pp 914–922
4. Zhang W (1994) Prediction of radio waves attenuation due to melting layer of precipitation. IEEE Trans Antennas Propag 42:492–500
5. Panagopoulos AD, Arapoglou PM, Cottis PG (2004) Satellite communications at KU, KA, and V bands: propagation impairments and mitigation techniques. IEEE Commun Surveys Tutorials 6:2–14
6. Dissanayake A, Allnutt J, Haidara F (1997) A prediction model that combines rain attenuation and other propagation impairments along earth-satellite paths. IEEE Trans Antennas Propag 45(10):1546–1558
7. Capsoni C (1987) Data and theory for a new model of the horizontal structure of rain cells for propagation applications. Radio Sci 22(3):395–404
8. Arbesser-Rastburg BR, Paraboni A (1997) European research on Ka-band slant path propagation. Proc IEEE 85:843–852
9. Carter J (2019) https://www.techradar.com/news/everything-you-need-to-know-about-spacexs-starlink-plans-for-space-internet
10. Kanellopoulos JD, Panagopoulos A (2001) Ice crystals and raindrop canting angle affecting the performance of a satellite system suffering from differential rain attenuation and cross-polarization. Radio Sci 36(5):927–940

11. Livieratos SN, Cottis PG (2001) Availability and performance of single/multiple site diversity satellite systems under rain fades. Eur Trans Telecommun 12(1):55–65
12. Fortuny J (1992) Satellite constellations for a global personal communications system at LBand. P. O. Box 299
13. Panagopoulos AD, Kanellopoulos JD (2003) Statistics of differential rain attenuation on converging terrestrial propagation paths. IEEE Trans Antennas Propag 51(9):2514–2517
14. Castanet L (2001) Comparison of various methods for combining propagation effects and predicting loss in low-availability systems in the 20–50 GHz frequency range. J Satell Commun 19:317–334

CoRaSat: A Marvel Satellite Technology with Bountiful Benefits of Cognitive Radio

Indu Bala and Samiya Majid Baba

Abstract Like many other wireless communication systems, the spectrum of satellite communication is also overcrowding day by day due to the fast adoption of multimedia applications, on-demand broadband communication, and related interactive services. Therefore, new communication paradigms are being explored for optimum spectrum utilization. In this regard, cognitive satellite communication has recently gained much interest in the research community. In this paper, the basic notion of satellite cognitive-communication and its practical implementation of hybrid/dual cognitive satellite systems are reviewed. Furthermore, a systematic review of various techniques for spectrum access is presented for dual satellite systems. In addition to that, several challenges associated with the applicability and deployment of cognitive radio over SatCom in the identified scenarios are presented and concluded the work by providing insights into the various open research issues in this domain.

Keywords Cognitive radio · Satellite communication · Interference · Spectrum access · Regulation · Standardization

1 Introduction

Recently, the demand for the wireless spectrum has increased significantly due to the faster adoption of high-speed internet-based interactive multimedia applications by people around the world [1, 2]. Satellite communication (SatCom) is the key enabling technology to fulfill the dire need for high-speed broadband access anytime, anywhere by 2025 due to large footprints and reachability to the remote geographical areas where the deployment of other terrestrial networks is infeasible. However, the main bottleneck of SatCom is improper spectrum utilization [3]. Due to the existing static spectrum allocation policies in use, most of the satellite frequency bands have already been assigned to the various services as shown in Table 1.

I. Bala (✉) · S. M. Baba
SEEE, Lovely Professional University, Jalandhar, Punjab, India
e-mail: i.rana80@gmail.com

D. Singh et al. (eds.), *Computer Aided Constellation Management and Communication Satellites*, Lecture Notes in Electrical Engineering 987,
https://doi.org/10.1007/978-981-19-8555-3_19

Table 1 Services offered by various satellite frequency band

Frequency band	Frequency range (GHz)	Bandwidth (GHz)	Type of service
C	4–8	4	Fixed satellite services
K	18–26.5	8.5	Fixed and broadcast satellite services
L	1–2	1	Mobile satellite services
S	2–4	2	Mobile satellite services, NASA, deep space research
Ku	12.5–18	5.5	Broadband services to vehicular users
Ka	26.5–40	13.5	Broadband services to vehicular users

It has been reported that the average spectrum occupancy rate of the pre-allocated spectrum bands is varying between 15 and 85% only [4]. To improve spectrum utilization, cognitive radio (CR) technology is evolving at a very fast pace for the coexistence with heterogeneous wireless networks [5]. CR is intelligent radio that learns from its surrounding environment and reconfigures its transmission parameters to provide the best quality of service to the end-users. It allows sharing of the spectrum dynamically by permitting unlicensed users access to the spectrum opportunistically on a non-interfering basis. Thus, the technology has the potential to improve spectrum utilization pre-allocated to terrestrial networks [6]. However, limited attention has been paid to the benefits that CR technology can bring to the satellite communication (SatCom) domain.

2 Spectrum Access Techniques Using CR

The various spectrum access techniques that have been discussed in the literature for cognitive radio can be classified as (i) interweave, (ii) underlay, (iii) overlay, and (iv) database (DB) related techniques. The interweave approach allows secondary users (SUs) to transmit the data on finding licensed primary users (PUs) absent from a specific band. The underlay techniques allow SUs to transmit the data along with licensed user-provided interference is not increasing to the licensed used beyond permissible values. The overlay technique mitigates interference using advanced level coding like dirty paper coding and improved transmission strategies [7], while the DB technique is a centralized architecture-based scheme in which the CR terminals refer database to identify unoccupied channels for data transmission.

While integrating the CR technology into the SatComs, the following aspects must consider as.

- Polarization and elevation angles can provide an extra degree of freedom to enable the integration of satellite and other wireless networks.
- The uplink transmissions can be done with high elevation angles to avoid interference with/from other wireless networks.

- Due to limited onboard power to satellites in space, interference-free transmission is inevitable.
- The narrowband spectrum sensing used in terrestrial CR systems is not applicable to detect the Ku/Ka-band signal.
- Highly directional fixed satellite terminals can be used for coexistence with other wireless technologies to meet end-user data requirements.

3 Deployment Scenarios for CR-Based Satellites

The significant literature is available on CR-based satellite communication which can be broadly classified into two categories, namely (i) hybrid CR-based SatComs and (ii) dual CR-based SatComs.

3.1 Hybrid CR-Based SatComs

The satellite and terrestrial networks can coexist to enhance spectral efficiency as shown in Fig. 1. The hybrid scenario could consist of multiple users or a single-user case as shown in Fig. 1a, b, respectively. In a multiuser scenario, S or C band satellites can coexist with other wireless networks. The licensed spectrum of satellite communication can be shared with other wireless network users using an appropriate spectrum access technique or vice versa. Whereas, in the single-user case, Ka and FSS satellite systems can share the spectrum allocated to the microwave fixed service (FS) [8].

3.2 Dual CR-Based SatComs

As shown in Fig. 2, dual CR-based geostationary earth orbit (GEO) satellites coexistence in Ka-band is shown where the primary user is a GEO satellite and the secondary user can be GEO or LEO/MEO satellite.

4 Challenges to Integrate CR Technology into Satellite Communication

In this section, some of the challenges are highlighted while integrating the technology into SatCom.

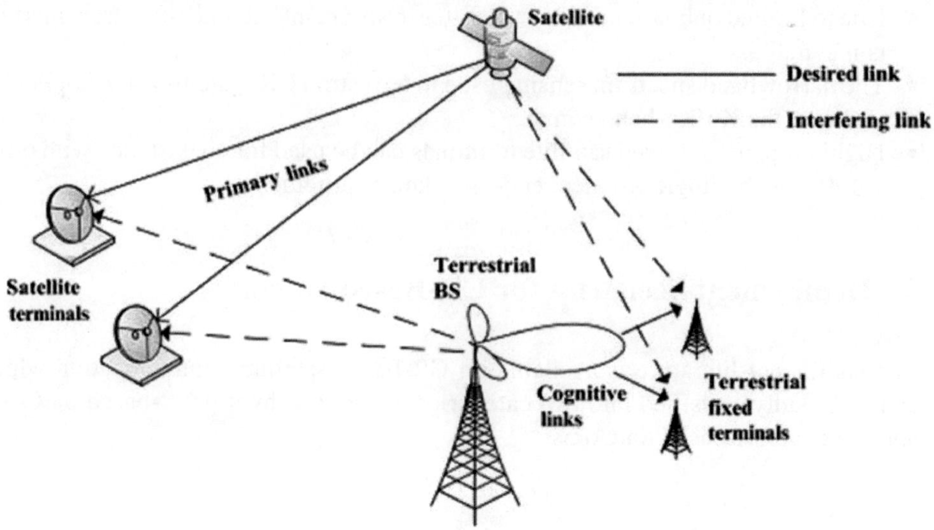

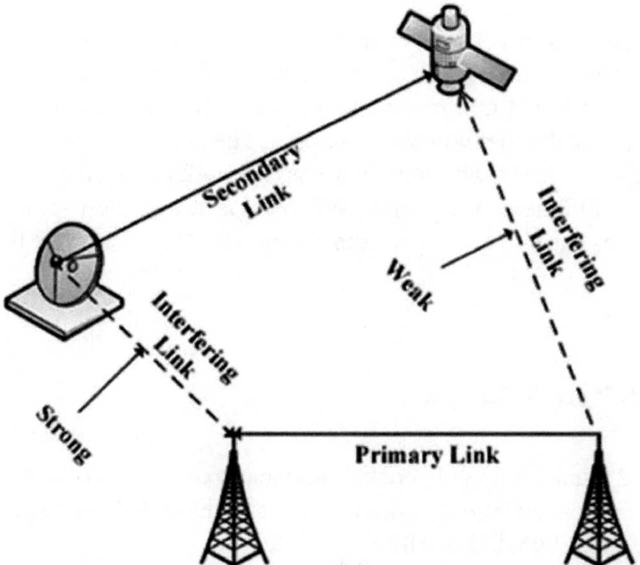

Fig. 1 Hybrid CR-based SatComs **a** Multiuser scenario, **b** Single user scenario

4.1 Regulatory Challenges

To enhance spectrum utilization, the fundamental challenge to the SatCom industry is to improve end-to-end connectivity, enhanced throughput, less transmission costs, interference management, etc. There is a significant development in this regard to

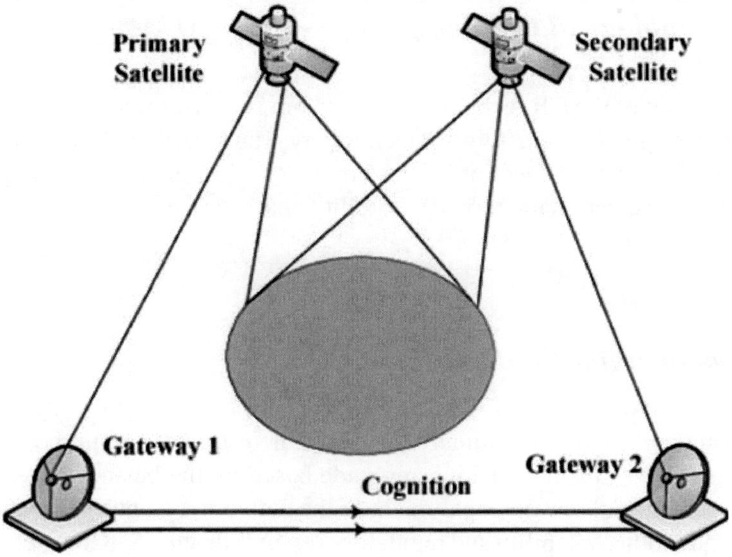

Fig. 2 Dual CR-based SatComs scenario

the terrestrial through a dynamic spectrum access approach. However, spectrum allocation for SatCom still follows traditional static approaches. Thus, there exists a dire need for the development of a regulatory framework to exploit the benefit of both technologies together. The regulatory authorities should encourage the usage of non-coordinated equipment and must also address the regulatory and licensing issue for the development of a real-time transmission coordination system [9].

4.2 Standardization Challenges

Dynamic spectrum sharing is already in use for terrestrial networks at 5 GHz. The use of CR in SatCom systems can enhance spectrum efficiency manifolds. However, it requires that the relevant standardization organizations investigate the impacts of hybrid architecture on the existing SatCom systems. In this regard, many standardization activities are happening around the globe such as the ITU-R activities on cognitive radio systems (CRS), the IEEE 1900 series of standards on dynamic spectrum access (DSA), the IEEE 802.22 standards on wireless regional area networks (WRANs) operating on TV white spaces, as well as the ETSI activities on CRS in TV white space bands and RRS [10].

4.3 Technological Challenges

The integration of the CR technology to SatCom scenarios requires a thorough revision and redesign of the existing CR technology. Many issues need to be addressed by CR techniques such as spectrum access methods, system architecture, propagation delay, jitter, receiver characteristics, satellite characteristics, wide-coverage, feeder link design, power levels, etc.

4.4 Commercial Challenges

SatCom networks require millions of dollars of investment, thus proper planning and deployment. Investment decisions are made based on the business case and market scenario. After a successful deployment of the network, revenue generation depends on the bandwidth availability and regulatory regimes that the SatCom network serves. The business challenges involve the efficient use of existing unused frequency bands by a secondary user equipment. The development of such new techniques, equipment, and infrastructures needs investment. Moreover, the services offered to the end-users must be delivered at an affordable cost. Another challenge that will be faced by investors is interference mitigation as it affects the overall revenue. Interference is a major issue for every satellite operator as it is affecting the core business of everyone [11].

5 Future Scope and Recommendations

The amalgamation of CR and satellite communication is a new concept and need time to get mature. In this section, some recommendations are provided and some issues are highlighted that need open discussions. These are.

The channel modeling for space is different from the terrestrial networks. Thus, the interference modeling must consider the cumulative interference from inter-satellite and terrestrial networks. And, the interference impact on the licensed users in context to the cognitive SatComs can also be investigated in the future.

The most of the existing literature, authors have considered the scenario comprised of a single licensed user. However, in practical scenarios, there exist many unlicensed users in the targeted frequency band and opportunistic access to the frequency band. Thus, while calculating the sum-rate capacity of such a system, due consideration must be given to this aspect as well. Moreover, more sophisticated coexistence techniques must be developed to accommodate multiple primary as well as secondary systems. CR technology is a well-investigated topic for terrestrial networks. Such systems opportunistically exploit the available frequency bands. The narrowband spectrum sensing is mandatory for such networks. However, when this technique is employed on satellite systems, wideband sensing techniques must be used. There is significant scope for developing new resource allocation techniques for SatCom systems.

6 Conclusion

The amalgamation of cognitive radio with satellites is an interesting concept and can be proven key in enabling technology to use the frequency bands efficiently to satisfy the needs of end-users with adequate quality of service. In this paper, we have explored the concept and reviewed some of the possible deployment scenarios for hybrid CoRaSat communication. Various deployment challenges have been highlighted, and recommendations are given for future research in this field.

References

1. Bala I, Bhamrah MS, Singh G (2017) Capacity in fading environment based on soft sensing information under spectrum sharing constraints. Wireless Netw 23(2):519–531
2. Bala I, Bhamrah MS, Singh G (2017) Rate and power optimization under received-power constraints for opportunistic spectrum-sharing communication. Wireless Pers Commun 96(4):5667–5685
3. Sharma SK, Chatzinotas SYMEON, Ottersten BJORN (2013) Cognitive radio techniques for satellite communication systems. In: 2013 IEEE 78th vehicular technology conference (VTC Fall), , September 2013. IEEE, pp 1–5
4. Bala I, Bhamrah MS, Singh G (2019) Investigation on outage capacity of spectrum sharing system using CSI and SSI under received power constraints. Wireless Netw 25(3):1047–1056
5. Bala I, Ahuja K (2021) Energy-efficient framework for throughput enhancement of cognitive radio network. Int J Commun Syst 34(13):e4918
6. Bala I, Ahuja K (2021) Energy-efficient framework for throughput enhancement of cognitive radio network by exploiting transmission mode diversity. J Ambient Intell Human Comput 1–18
7. Ahn DS, Kim HW, Ahn J, Park DC (2011) Integrated/hybrid satellite and terrestrial networks for satellite IMT-Advanced services. Int J Satell Commun Network 29(3):269–282
8. Kandeepan S, De Nardis L, Di Benedetto MG, Guidotti A, Corazza GE (2010) Cognitive satellite terrestrial radios. In: 2010 IEEE global telecommunications conference GLOBECOM 2010, December 2010. IEEE, pp 1–6
9. Filin S, Harada H, Murakami H, Ishizu K (2011) International standardization of cognitive radio systems. IEEE Commun Mag 49(3):82–89
10. Sharma SK, Chatzinotas S, Ottersten B (2012) Satellite cognitive communications: Interference modeling and techniques selection. In: 2012 6th Advanced satellite multimedia systems conference (ASMS) and 12th signal processing for space communications workshop (SPSC), September 2012. IEEE, pp 111–118
11. Vanelli-Coralli A, Guidotti A, Tarchi D, Chatzinotas S, Maleki S, Sharma SK, Liolis K (2015) Cognitive radio scenarios for satellite communications: the CoRaSat project. In: Cooperative and cognitive satellite systems. Academic Press, pp 303–336

Sub-banding-Based Digital Beamforming for Transmission of Wideband Signals

Priyanka Das, K. R. Yogesh Prasad, L. Suvarna, N. Ramalakshmi, and D. Venkataramana

Abstract The article discusses an efficient method for digital beamforming toward transmission of signals having large bandwidth. Conventional digital beamforming techniques offer hardware efficient solution when compared to active arrays implemented using analog microwave phase-shifter modules. However, these techniques exhibit an inherent limitation while signals having large bandwidth are under consideration. The article discusses a methodology to overcome this constraint. By resolving the wideband signal into its constituent frequency sub-bands and subsequently computing the phase-shifts based on the center frequency of the sub-bands, it is shown that the radiation pattern of the active antenna can be improved. The article compares the results obtained by simulating a wideband signal transmission using a uniform linear array adopting the conventional approach as well as sub-banding. The approach has several applications including space-based data transmission.

Keywords Data transmission · Phased array antenna · Digital beamforming · Sub-banding · Discrete Fourier transform · Radiation pattern

1 Introduction

The demand for high data rate transmission is steadily on the rise for applications ranging from terrestrial mobile networks to space-based payload data transmission. Several aspects such as frequency of transmission, modulation scheme, sizing of power amplifier, antennas, etc., need to be taken into consideration while configuring a communication system for a given application [1–4]. Figure 1 shows the elements of a typical communication system [5].

Higher data rates call for higher carrier frequency while bandwidth constraints of the channel influence the choice of modulation. The RF power required to be generated by the power amplifier is closely linked to the data rate, chosen modulation scheme, and the antenna system employed. While wide beamwidth antennas help in

P. Das (✉) · K. R. Yogesh Prasad · L. Suvarna · N. Ramalakshmi · D. Venkataramana
U R Rao Satellite Centre, ISRO, Bangalore 560017, India
e-mail: priyanka.isro@gmail.com

D. Singh et al. (eds.), *Computer Aided Constellation Management and Communication Satellites*, Lecture Notes in Electrical Engineering 987,
https://doi.org/10.1007/978-981-19-8555-3_20

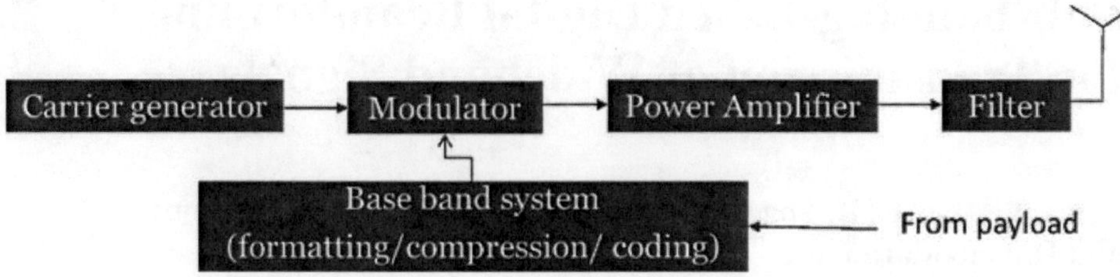

Fig. 1 Block diagram of a typical data transmission system

increasing the area of coverage, they can place impractical demands on the amplifiers. On the other hand, a high gain antenna system can reduce the RF power requirement at the cost of reduced coverage. The contradictory requirements are often met by employing a phased array antenna (PAA) system which offers high gain as well as the capability to steer the beam in any desired direction, thus increasing the area of coverage. Digital beam forming (DBF) techniques are commonly used in conjunction with PAA [6–8]. DBF technique for wideband signal calls for processing of the signal by sub-band decomposition [9].

In the sub-band decomposition process, the wideband signal is divided into several frequency bands and narrow band DBF process is applied on these sub-bands individually. Another method of processing the wideband signal is to process them using tapped delay lines as done in FIR filters and then converge them in the desired direction. Adaptive approaches to reduce computational complexity and to achieve faster convergence suggest combining the spatial and temporal beam forming approaches. For ultra-large bandwidth signal such as beamforming for hybrid MIMO communication application, some mixed mode of optimization is required such as virtual sub-array and true-time-delay-based hybrid approach to get a uniform response as well as avoid beam squinting [10].

The article focuses on the design aspects of phased array antenna system required for supporting wideband signal for data transmission application. Simplified approach to meet the requirements of this application is discussed. Section 2 of the article provides an introduction to the concept of phased array antenna and discusses the popular approaches for beamforming using PAA and their limitations. Approach for supporting high bandwidth signals based on sub-banding is discussed in Sect. 3. Simulations demonstrating the advantages of sub-banding approach are presented in Sect. 4. Scope for future work in this area is brought out in Sect. 5.

2 Phased Array Antenna for Data Transmission

2.1 Analog Beamforming

An antenna array consists of a group of smaller radiating elements which collectively operate to form an antenna of a larger aperture. The signals to the individual radiating elements are fed so as to interfere constructively in the desired direction. For an array to act as a transmit antenna, the phase distribution of signals is controlled so that the wavefront, which is the locus of points of signals having the same phase, is formed in the direction perpendicular to the desired direction of propagation. Figure 2 shows the block schematic of a typical phased array antenna employing analog beamforming [11] and the process of wavefront formation is depicted in Fig. 3.

By controlling the phase-shifts to the individual elements, the direction of propagation can be controlled as shown in Fig. 4a, b.

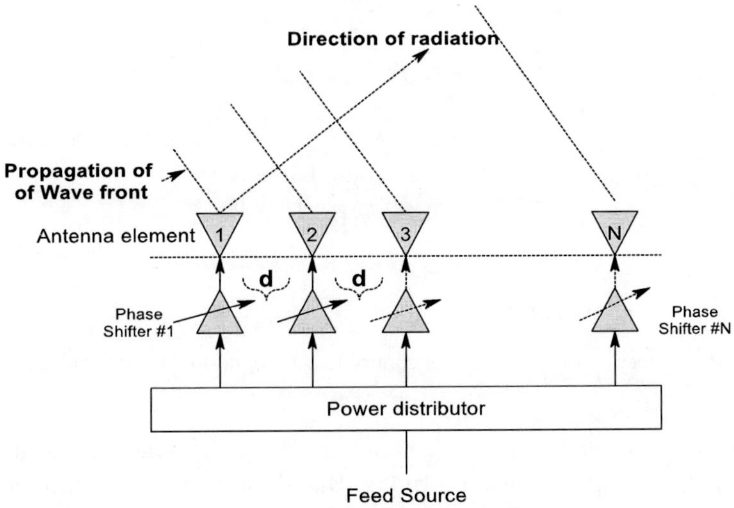

Fig. 2 Block schematic of phased array antenna employing analog beamforming

Fig. 3 Formation of wavefronts formed by equiphase signals in linear and circular arrays

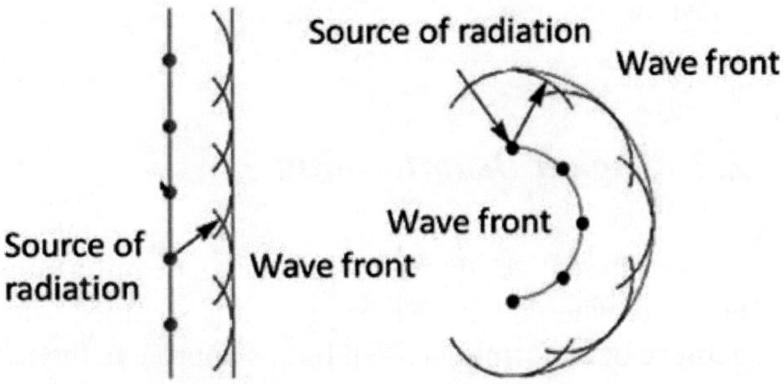

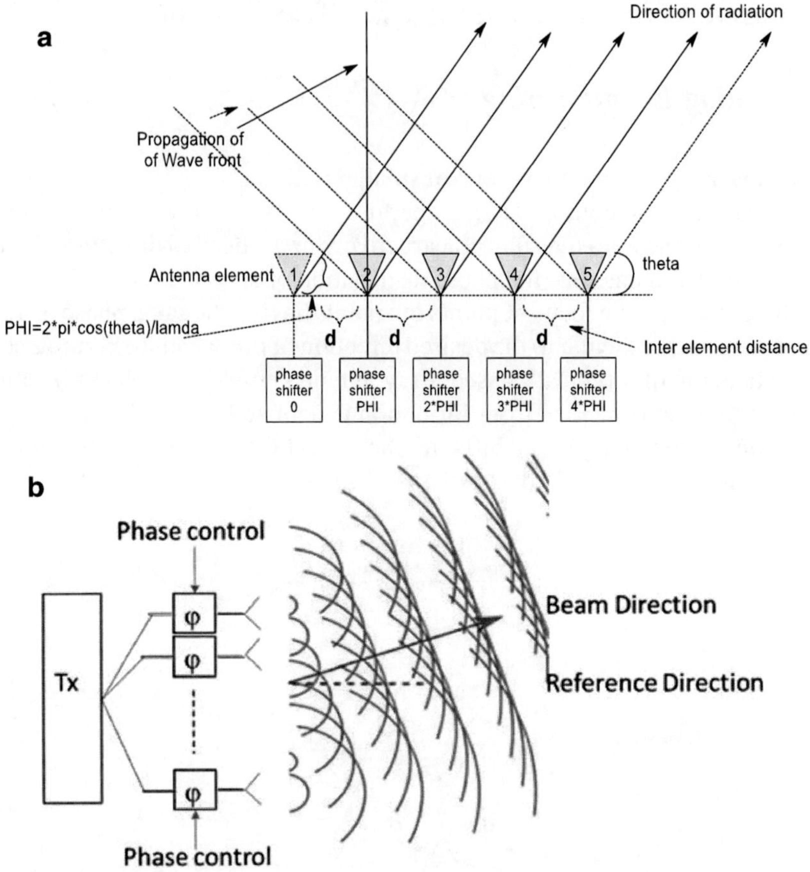

Fig. 4 **a** Beam steering in PAA. **b** Phase control to achieve desired beamforming

Digitally controlled phase-shifters are commercially available to enable control of signal phase-shifts for beam forming. Based on the array configuration and the desired direction of beam forming, the necessary phase-shifts are computed in real time by an electronic PAA controller. However, this approach would call for multiple phase-shifter modules which in turn increases the hardware and the associated power consumption.

2.2 Digital Beamforming

As an alternative approach to analog beamforming, digital beamforming may be adopted wherein the RF signals are sampled, stored, and digitally processed to achieve beamforming [6–8]. In this approach, the desired amplitude and phase transformations are carried out by multiplication of the signal samples with appropriate complex numbers. The set of complex numbers for the transformations vary based on

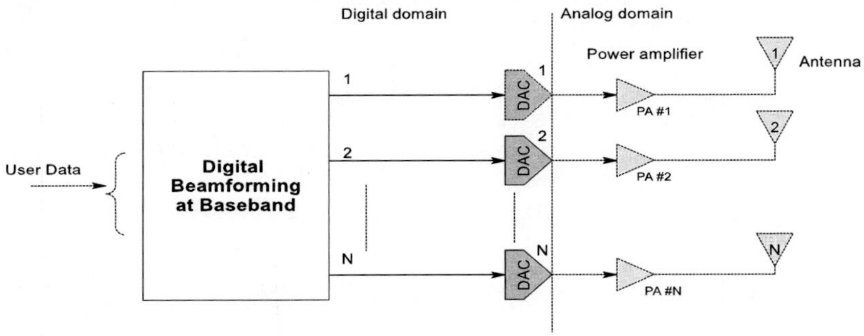

Fig. 5 Phase control to achieve beam steering in a PAA

the desired direction of beam forming and in turn based on the phase-shift required to compensate the differential path lengths traversed by the signals to reach the far-field point.

Thus, it is possible to generate multiple beams using the same set of signal samples. This approach offers advantages in terms of reduced hardware as the phase-shifting is accomplished by digital signal processing instead of phase-shifter modules. Figure 5 indicates the block schematic of the RF section of a typical digital beamforming-based PAA.

2.3 Limitations of the Conventional Digital Beamforming Approach

While the digital beamforming approach offers several advantages, it has some inherent limitations when it comes to applications involving large bandwidth signals. The computation of phase-shifts required for beam steering is typically carried out for the center frequency of the signal bandwidth and applied to the composite signal composed of various spectral components. This conventional approach is referred to as narrowband processing for digital beamforming.

When the signal bandwidth is large, the phase-shift computed for the center frequency does not compensate the phase-shift actually experienced by the frequency components that are spectrally distanced from the center frequency. This leads to improper phasing of such components at far-field points where the signal is received, which in turn degrades the radiation pattern and fidelity of the signal.

In order to study the effects of narrowband processing on wideband signals when transmitted by digital beamforming, the radiation pattern of a QPSK modulated signal centered at 8 GHz and spread over 7–9 GHz is considered. The radiation pattern generated by narrowband processing at 8 GHz and also at band edges of 7 and 9 GHz are plotted. Figure 6 represents the radiation pattern of the PAA formed at commanded elevation angle of 30°, for different frequency components. It may be noted from Fig. 6 that the components away from the center frequency are considerably affected.

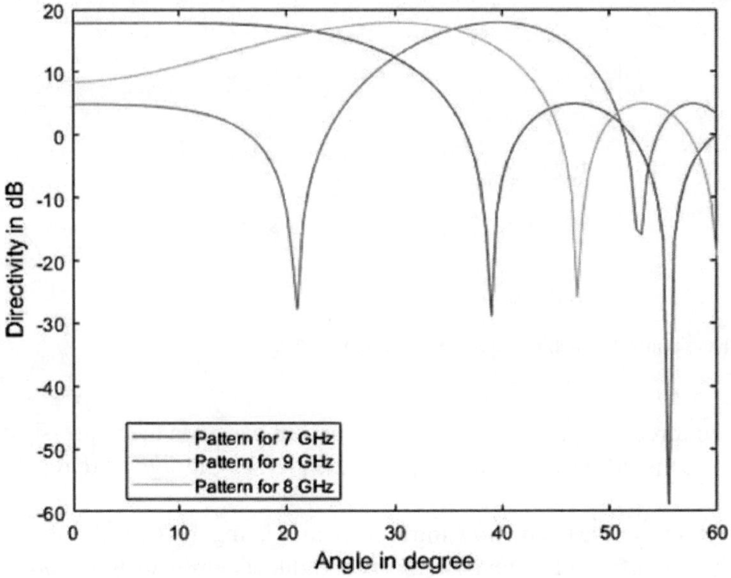

Fig. 6 Comparison of PAA radiation pattern plots for 7, 8, 9 GHz frequency components formed by conventional narrowband beamforming process for beam along (30°, 0°)

3 Sub-banding Approach for Digital Beamforming of High Bandwidth Signals

In order to overcome the limitations of conventional narrowband processing for digital beamforming, signal processing approach called sub-band processing is adopted. The signal is decomposed into multiple sub-bands and narrowband processing is applied to each of these sub-bands [9]. Processed signals from all the sub-bands are combined to form the output signal.

While adopting sub-banding, the signal to be transmitted over PAA is converted to frequency domain by the use of fast Fourier transform (FFT). The number of points considered for the transform determines the number of frequency bins into which the composite wideband signal is decomposed. The center frequency of each bin is chosen to carry out narrow band beamforming for frequency components belonging to the chosen bin. This approach restricts the spectral distance between the frequency chosen for phase compensation and the band-edge frequencies. Phase compensation for path length difference is then carried out by multiplication of the Fourier components with computed complex exponentials which determine the amplitude and phase transformations for a given frequency bin. All the transformed frequency components reaching a given radiating element are combined after being translated to time domain by applying inverse fast Fourier transform. This results in radiation of signals, which interfere constructively along the desired beam direction, irrespective of the bandwidth of the signal under consideration as long as FFT is carried out such that the spectral width of the bins is limited.

The block schematic capturing the functional blocks of sub-banding approach for digital beamforming is shown in Fig. 7. The output of quadrature modulator represents the wideband signal to be transmitted. N-samples of this signal, sampled at Ts = $1/F_S$, are processed together to obtain N-point FFT of the signal. The output of FFT comprises of N frequency points that are separated by 1/(N*Ts), which forms the frequency resolution for subsequent processing. If a linear array comprises of M elements, the N-point FFT is replicated M times. The N-points of FFT map the frequencies ranging from 0 to Fs on to N sub-bands whose center frequencies are given by

$$F_{\text{sub},n} = F_s/(2N) + (n - 1) * F_s/N \quad \text{for } n = 1 \text{ to } N$$

Phase compensation is done for all the M radiating elements and for all N number of sub-bands, based on their location in the array. The phase compensated signal reaching each of the radiating elements is transferred back into time domain by applying inverse Fourier transform before being fed to the respective radiating elements. This approach ensures that the maximum difference between the frequency

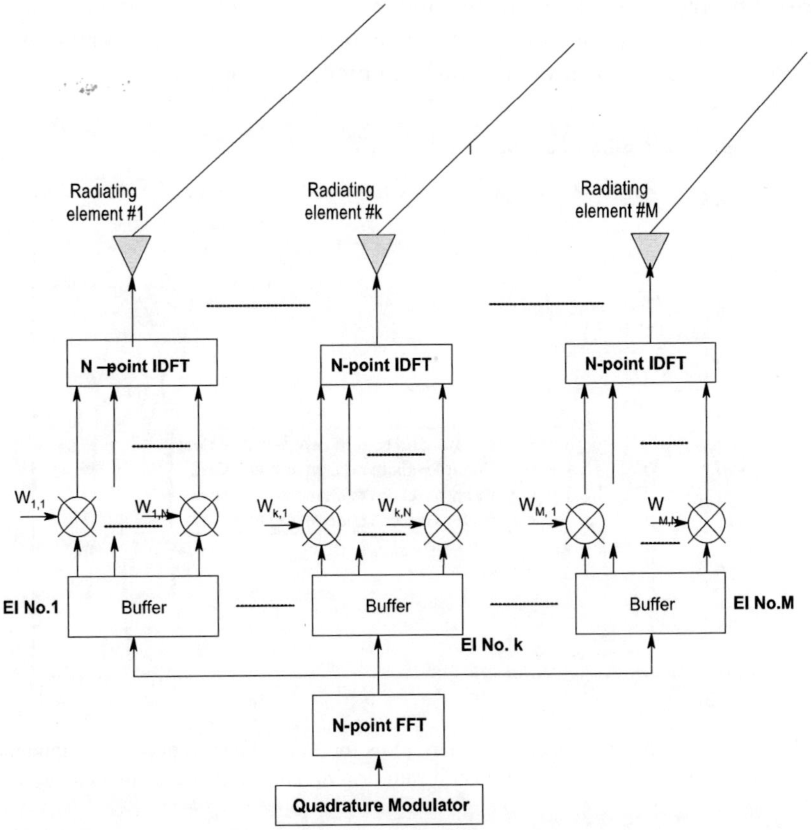

Fig. 7 Block schematic of sub-band processing for digital beamforming for a transmit application

considered for phase compensation and the radiated signal is limited to half the width of the sub-band, which is much smaller than half the bandwidth of the signal seen in narrowband processing approach.

4 Simulated Patterns and Discussion

In order to compare the narrow band and sub-band processing approaches, the radiation patterns generated by the two approaches are plotted for the signal characteristics considered in Sect. 3.

It may be noted from Fig. 8 that the directivity along the commanded direction (0° elevation) obtained by narrowband processing as well as sub-banding are the same for center frequency component at 8 GHz. However, the limitation of narrowband processing becomes evident for the frequency components that are away from the center frequency as seen by the drop in directivity for the 7 and 9 GHz components subjected to narrowband processing. The pattern formed by sub-banding is independent of the frequency of the transmitted component.

Plots in Figs. 9 and 10 indicate similar trends in radiation plots, commanded to elevation angles of 30° and 60°, showcasing the advantages of sub-banding approach for wideband signals over the narrowband processing approach.

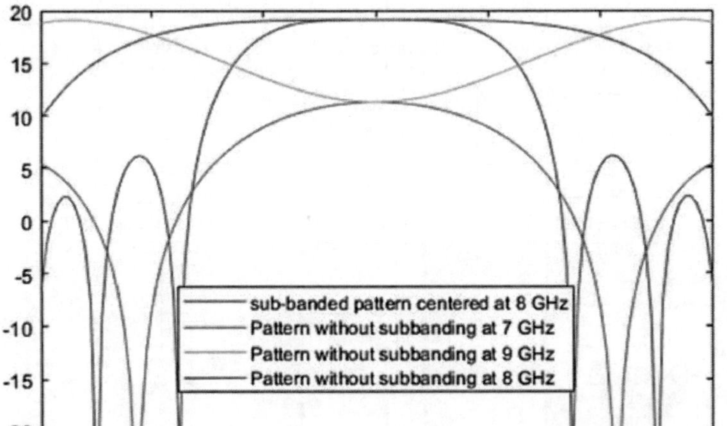

Fig. 8 Comparison of PAA radiation pattern plots for 7, 8, 9 GHz frequency components formed by conventional narrowband and sub-band beamforming processes for beam along (0°, 0°)

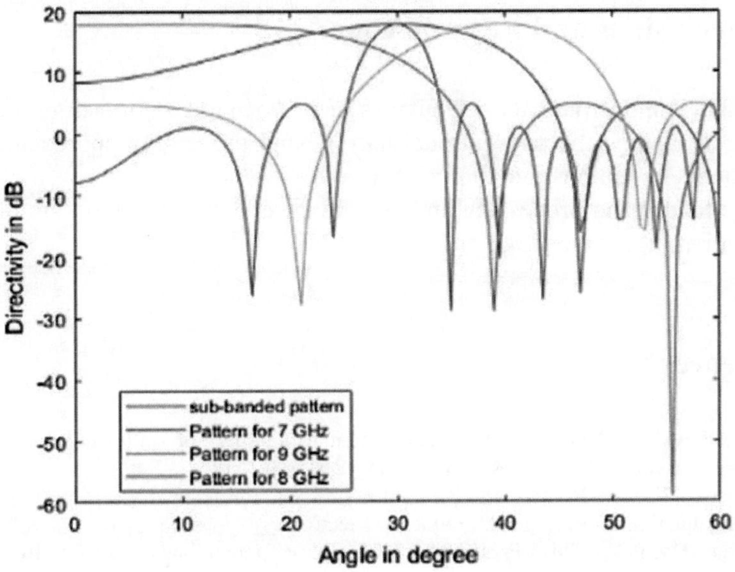

Fig. 9 Comparison of PAA radiation pattern plots for 7, 8, 9 GHz frequency components formed by conventional narrowband and sub-band beamforming processes for beam along (30°, 0°)

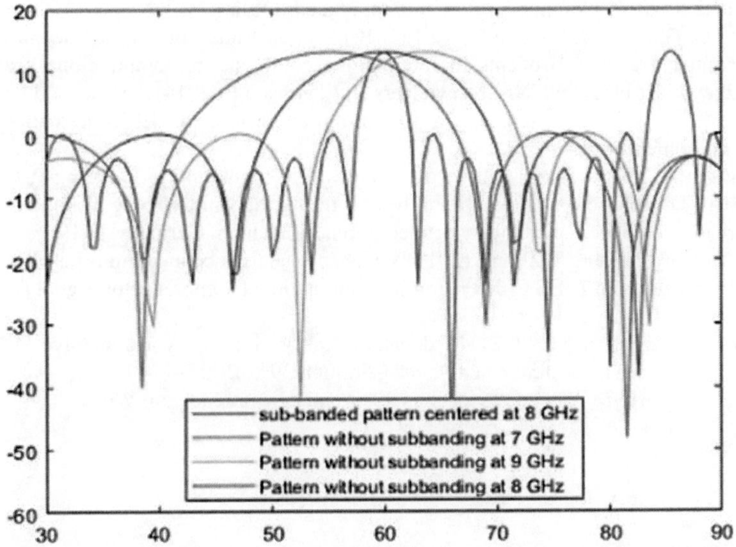

Fig. 10 Comparison of PAA radiation pattern plots for 7, 8, 9 GHz frequency components formed by conventional narrowband and sub-band beamforming processes for beam along (60°, 0°)

5 Conclusions and Future Scope

The article demonstrates the advantages of sub-banding approach to digital beamforming as against the conventional narrowband processing approach. The radiation patterns to demonstrate the same have been plotted. Hardware implementation aspects of sub-band processing and performance characterization are planned as a continuation of this work.

References

1. Majeed RM, Mclane PJ (1997) Modulation techniques for on-board processing satellite communications. IEEE Trans Commun 45(12):1508–1512
2. Ramana DV, Jolie R, Rao VS, Pal S (2008) A satellite telemetry transmitting system with pre-modulation filtering. High Frequency Electronics
3. Ramana DV, Pal S, Shiva Prasad AV (2005) Space applications for input pulse shaping filter. IEEE Aerospace Electron Syst Mag
4. Addabbo P, Antonacchio F, Beltramonte T, di Bisceglie M, Gerace F, Giangregorio G, Ullo SL (2014) A review of spectrally efficient modulations for earth observation data downlink. IEEE Metrology for Aerospace (MetroAeroSpace), Benevento, Italy
5. Haykins S (2007) Communication system, 2nd edn. Wiley
6. Liang G, Gong W, Jia B, Liu H, Yu J (2010) Demonstration of a digital beamforming (DBF) transmitter array with 16 beams. In: International conference on computational problem-solving
7. Raczkowki K, De Raedt WL, Nauwelaers B, Wambacq P (2014) A wideband beamformer for a phased-array 60 GHz receiver in 40 nm digital CMOS. In: IEEE international solid-state circuits conference
8. Sangwan A, Ma R, Wang B, Kim K, Parsons K, Toshia, Akino K, Wang P, Orlik P, Teo K (2019) CDM-based 4-channel digital beamforming transmitter using a single DAC. In: IEEE International new circuits and system conference, Munich, Germany, 2019
9. Cao Y, Wang Y, Wang S, Zhou S (2019) Wideband subarray beamforming based on subband and decomposition. In: IEEE China summit & international conference on signal and information processing, China, 2019
10. Gao F, Wang B, Xing C (2021) Wideband beamforming for hybrid massive MIMO terahertz communication. IEEE J Selected Areas Commun 39(6):1725–1740
11. Balanis CA (1996) Antenna theory: analysis and design, 2nd edn. Wiley

Comparative Analysis of Secure QKD Protocols for Small Satellites Constellation

Hardeer Kaur and Jai Sukh Paul Singh

Abstract Nowadays, everyone uses Internet, do financial transactions and many other day-to-day activities. As the users are becoming more and more aware of the privacy, data security had become a prime concern. Quantum Internet inspires to be a possible solution. Quantum Internet uses quantum key distribution to provide security by generating symmetric keys for the data transfer. Newly developed technologies and manufacturing processes paved way for designing and development of small satellites. Small satellites can be launched in forms of clusters at a marginal cost of launching large communication satellites. Here, we review the basic concepts of quantum key distribution. Its application in satellite communication specifically using CubeSat satellites. We also touch upon the recent development in the field of satellite-based QKD systems. A brief on the major milestones and important small satellite mission is also presented. Further, we present a comparison study between the prominent QKD algorithms pertaining to their application in satellite-based QKD systems. We conclude the article with discussion on the applications, advantages and disadvantages of using QKD on satellite-based communication networks, further extending it from local area networks to worldwide Internet.

Keywords Quantum cryptography · Quantum internet · Quantum key distribution

H. Kaur (✉) · J. S. P. Singh
School of Electronics and Electrical Engineering, Lovely Professional University, Phagwara, Punjab, India
e-mail: hardeer_kaur@yahoo.com

J. S. P. Singh
Division of Research and Development, Centre of Space Research, Lovely Professional University, Phagwara, Punjab, India

Division of Research and Development, Department of Research Collaboration, Lovely Professional University, Phagwara, Punjab, India

D. Singh et al. (eds.), *Computer Aided Constellation Management and Communication Satellites*, Lecture Notes in Electrical Engineering 987,
https://doi.org/10.1007/978-981-19-8555-3_21

1 Introduction

Quantum literally means "how-much" in Latin. In addition, true nature to its name, quantum physics is how-much one what's to learn. Quantum physics finds its use in all modern day applications. We would like to bring reader's attention to the applications of quantum mechanics in the field of cryptography and further it's applications in the satellite communication. In quantum application (pertaining to the computing), we use QUBITS (or Quantum bits). Bits used in classical computers can have only one out of two values, i.e., either '0' or '1'. But when talking about quantum bit or QUBITS, they can attain values '0' or '1' or 'anything in between' as shown in Fig. 1.

With the recent developments in the quantum computing, one can be pretty sure that engineers will be able to develop a full-scale quantum computer in near future. As quantum computer can solve complex problems with ease, it finds it's applications in the fields of complex simulations like chemical reaction simulation, weather modeling, cryptography, etc.

In contemporary cryptography (such as RSA), we use complex mathematical calculations to generate a key which then is used to encode and decode the data. One can hack or guess the key using a simple brute force attack. In one such experiment, it was estimated that to brake commonly used 256 bit key using a 2.8 GHz computer with 4 cores will require 13 months. In contrast, if we have a quantum computer using Shor's algorithm [1] with sufficiently large number of QUBITS, such key can be guessed in just one sec. Similarly, Grover's algorithm can be used to find objects in an unsorted array [2]. Hence, once the quantum computers are available most of the current cryptographic will become redundant. Going forward two new branched have immerged in the field of cryptography.

Post Quantum Cryptography (PQC). These are a set of algorithms or procedures for cryptography, capable in tolerating an attack from quantum user. These algorithms primarily based on the classical computing procedures and does not use the principles of quantum computing. They have distinct advantage that they can be applied on the currently available networks.

Fig. 1 Classical bits and quantum bits

Quantum Cryptography or Quantum Key Distribution (QKD). Basically, these protocols use the basic concepts of quantum physics to generated key. As these keys are generated using quantum physics in contrast to mathematical expressions they are found to be virtually hack proof. Hence, sometimes referred to as "The Holy Grail" of cryptography.

1.1 Basic Principles of Quantum Physics

Quantum physics as we know is field of science dedicated to understand the properties of quantum particles. The basic concepts of quantum physics can be understood by following principles.

Quantization. This in basic terms is discretization of energy into smallest possible units (in a way that no further division is possible) Fig. 2.

Using the principle of quantization, we work on the basic particles, which exhibits quantum behaviors. Most commonly used quantum particles are photons, neutrons, electrons, protons, etc.

Uncertainty Principle. This principle defines that we cannot predict the position (or state) of any particle with 100% surety without physically measurement. One can only predict in terms of probability of finding the particle at a desired location Fig. 3.

The certainty of predicting the state increases with the increase in the size of the object. As in quantum application, we work with small particles the probability function is very important.

No Cloning. By no cloning principle, it is intended that one cannot impose a state on the quantum particle or in other words cannot create enforce a desired quantum state on the particle. We can definitely changes the state, but we do not have the control over the initial states of the particles. As we cannot enforce a state, we cannot make copies of the quantum particles.

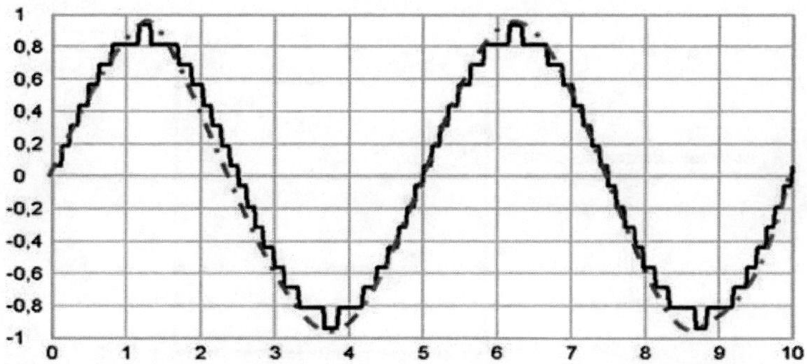

Fig. 2 Quantization of a light signal

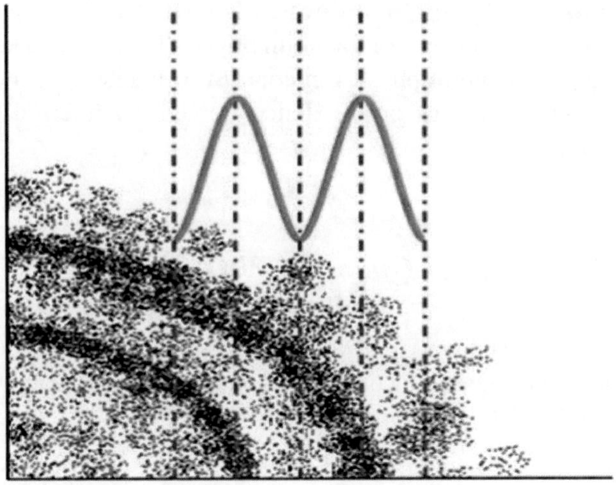

Fig. 3 Principle of uncertainty (probability of finding the electron)

Quantum Entanglement. Quantum particles are said to be in entangled state when by knowing the state of one particle the state of other particle is known. The state of both the partials collapse once the measurement is done on one of the particle. It can be better understood by the following example Fig. 4.

If we prepare a set entangled photon from a single input laser, then the total energy stored by the photons will be equal to the energy of the input laser pulse. If we measure the energy of one photon, we will defiantly know the energy of the second photon. Taking an example as shown in the above figure, considering the input beam is split using a beam splitter to create two entangled pair of photons. Assuming the photons are entangled in their spinning states, each particle will have equal and opposite spin to it. If one is spinning in clockwise direction, the other has to be anticlockwise direction.

Tunneling (no energy barrier). This principle helps the quantum particles to sometimes cross the energy barrier even while having energy less than that of the barrier itself.

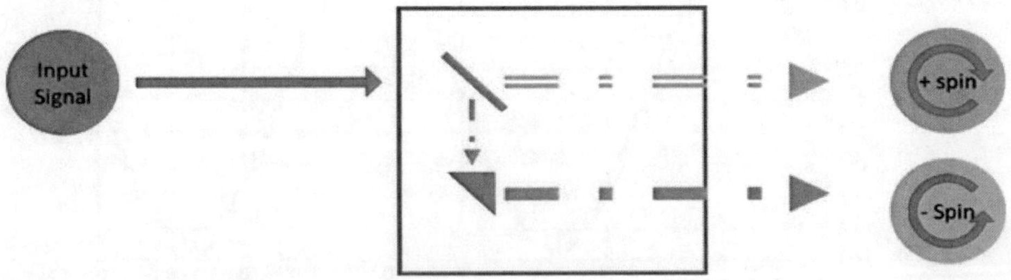

Fig. 4 Quantum entanglement

2 Quantum Cryptography

Quantum cryptography or quantum key distribution (QKD) provides users with secure key that can be used to encrypt and send data across a public channel. QKD uses principals to generate and distribute symmetric keys. Quantum cryptography term can be traced back to year 1970 when Stephen Wiesner wrote a conjugate code with an intent to solve the problem of counterfeit currency [3]. Later in year 1979, Bennett and Gilles Brassard developed a cryptographic system Crypto82 combining the concepts of public key cryptography and Wiesner's concepts. This system or algorithm was published in year 1984 with the name of BB84 [4]. It is worth noting that this was the first paper to actually give a terminology of "quantum cryptography". Going forward many QKD algorithms were developed, i.e., BB84, B92, SARG04, COW, etc., to name a few. Initially it was recommended to use one time pad for encoding the data, this was done to ensure that eve will not be able to predict the key. If the key is reused multiple times it allows eve to gain some understanding of the key. Later in the year 2014 Bennett proposed that the generated key can be reused multiple times provided that there is no eve is present in the system. He concluded that quantum systems can detect presence of eve with certainty and key pad need to be updated only once eve is detected [5].

Quantum key distribution algorithms can be broadly classified in the following categories:

Discrete Variable QKD (DV-QKD). Also known as prepare and measure protocols, these protocols typically use single photon source and detector. Most commonly used DV protocols are BB84, E91, BBM92, BB92, SSP, DPS, BB84 Decoy State, etc.

Continuously Variable QKD (CV-QKD). In these protocols, pulses of particles are used and encoding is done in amplitude and phase angles. Similarly, measurements are done using balanced homodyne detectors. COW, TF-QKD, MSZ96, etc., are some of the most commonly used protocols.

Device Independent QKD (DI-QKD). For quantum key distribution to be secure a basic requirement is to have highly secure nodes with perfect equipment. In short a utopian system, which is nearly impossible to maintain in practical scenario. Now researchers are developing a new set of protocols, which can allow users to communicate with each other using untrusted nodes in between. These new set of protocols are listed under device independent protocols. DI and MDI QKD algorithms are examples of DI-QKD algorithms.

Public Key Cryptography QKD. Here, a combination to two keys is used, i.e., public key and private key. In most basic form, users have their own private key, which they are going to utilize to communicate on a quantum channel. Alice (sender) uses her own private key to encode the photons and send the encoded photons to bob (receiver) through a quantum channel. Bob at his end takes his measurements using his own private key and send back the message to Alice. Alice once again can measure the message came from bob using her own private key to get access to the information send by bob.

Semi Quantum or Quantum enabled Key Distribution. These are special key distribution protocols, which uses a combination of classical as well as quantum mechanics. These protocols are developed based on an assumption that in near future not everyone will have devices which are quantum enabled. Hence, to attain a level of security, there is a need for a hybrid system where a key generation can be done between a quantum enabled device and another classical device.

One such example is to use quantum random number generators (QRNG) to generate random numbers for classical cryptographic algorithms. Many of the classical cryptographic algorithms use random numbers to generate keys. Hence, the security of the encryption is a function of randomness of the base random number used. For all practical purposes, a mathematical algorithm is used to generate random number. As it is based on a mathematical expression, the random numbers are not truly random [6, 7]. One can principles of quantum physics to easy generate random numbers which are truly random.

3 Quantum Communication with Satellite Links

Richard Hughes [8] and his team at LAS Alamos in 1990's first propose the use of satellite for quantum key distribution over large distances. Later on the idea was adopted By European Space Agency (ESA) conducted several feasibility studies for finding the potential of quantum communication with satellite links [9, 10]. In year 2004, ESA launches its project "QUEST" with an aim to conduct several experiments and demonstration of free-space QKD [11–13]. They were able to demonstrate key distribution at a distance of 144 km using weak coherent pulse decoy state [14] and with entanglement-based quantum states [11]. It is well know that with the use satellite, effective length of the single hopped link can be greatly extended. Using satellites with high quality optical link, the distance between the ground stations can be greatly improve. For any optical communication, either free-space link or optical cable link is used.

When using optical fiber links due to scattering effect in the optical fiber. This limits the maximum distance of repeater-less communication. PLOB upper bound can be expressed as [15]

$$-\log 2(1-\eta) \text{ or } 1.44\,\eta \text{ bits per link} \tag{1}$$

When using free-space link the upper limit is defined as

$$-\Delta \log 2(1-\eta) \text{ bits per link.} \tag{2}$$

Δ is used to compensate for the turbulence in environment [16].

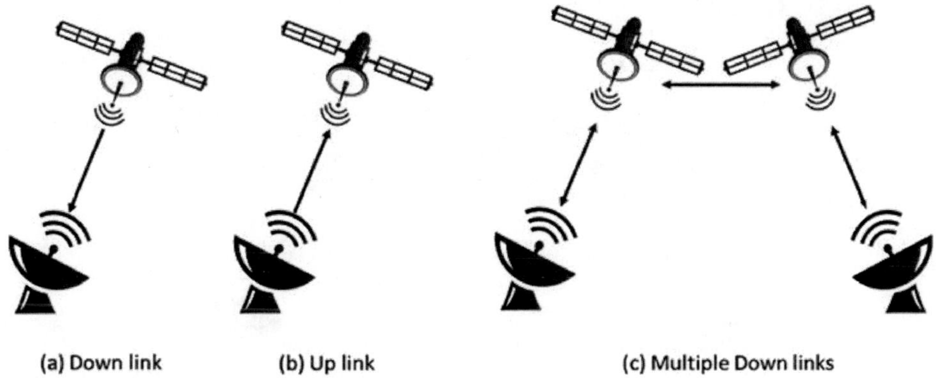

Fig. 5 Satellite communication configurations

Based on various QKD algorithms and their best possible implementations, multiple configurations of the satellite and ground stations can be thought of, some of the basic configurations are mentioned below:

Down link configuration. In downlink configuration as shown in Fig. 5a, the sender is placed at the satellite and the receiver is situated at the ground station. This configuration is suitable for the DV QKDs as the encoded photon are to be generated at only the satellite side.

Uplink configuration. It is the reverse of the downlink communication as shown in Fig. 5b. In this configuration, the more complex part link encoder, pointing devices are placed at the ground station and only detector is placed on the satellite. This configuration has its own advantages as for efficient working detectors require cryogenic cooling which can be achieved easily in space.

Satellite to multiple downlinks. This is in principle similar to the downlink configuration, but it was developed for the protocols which uses entanglement to distribute quantum key as shown in Fig. 5c. The source of entangled photon is placed on the satellite, and one photon each from entangled pair is transferred to both the intended participants.

Satellite Cluster configuration. The cluster formation the QKD is generated between the two-ground station and the photons are transferred through a series of satellite links as shown in Fig. 6. This configuration is thought of as the backbone of truly intercontinental quantum satellite networks. This configuration provides a multipath approach to the quantum network.

Deep Space Configuration. This configuration is designed for our future needs of securely communicating with the satellites used for deep space missions as shown in Fig. 7. Were satellites can communicate with each other or signal can travel from ground station to a deep space satellite using in between satellites as trusted repeaters.

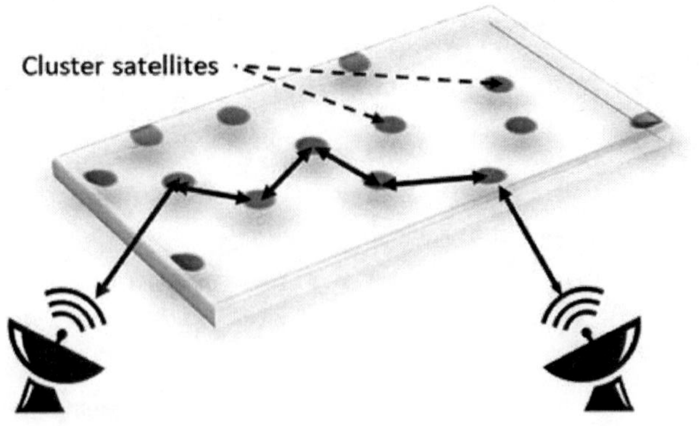

Fig. 6 Satellite cluster configuration

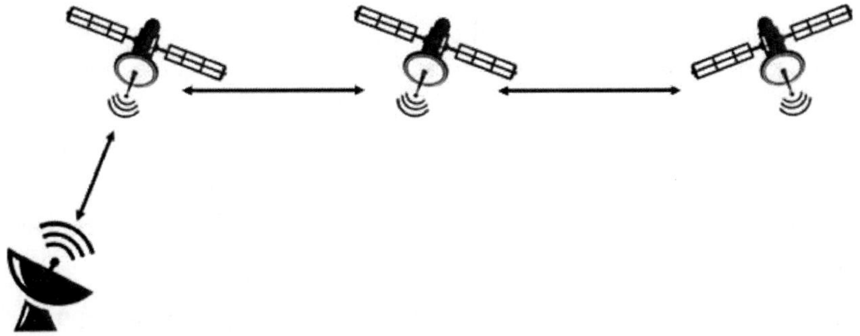

Fig. 7 Deep space configuration

3.1 Challenges of Using Optical Channels for QKD Generations

Everyone believes the satellite communication is necessary but to make it a reality we need to understand first, the challenges and limitation. One of the most important limitations of governing factor in the successful implementation of any QKD algorithm is bit error rate. This bit error rate is proportional to the errors generated while the signal is transferred through a quantum channel (most commonly used is optical link). The amplitude of error or loss depends upon the properties of the channel used. Most commonly used quantum particles are photons. Optical systems largely use either optical cables or free-space links. Optical cables being flexible has a distinct advantage for metropolitan networks. Sender and receiver are not required to be in straight line for communicating. However, optical fibers have issues like scattering and absorption. Whereas while using free-space communication the losses due to scattering are less but user has to be aligned in a straight line for the communication to happen. This channel has an advantage that it can be resolved at larger distances. It is most suitable for satellite communication. Some of the prominent limitations are:

Scattering in Optical cable. Optical cables although very effective and commonly used in classical communication and most of the technology required to build networks based on optical fibers are matured. Having said that the things are quite different when using it for quantum communication. As we cannot use repeaters and amplifiers small error or scattering leads to great reduction in either key generation rate or the distance for the effective key distribution. The upper bound limit for QKD distribution is defined by

$$\log_2(1-\eta)\ \text{bits per link.} \tag{3}$$

Transmission losses in optical beam. Beers law is used to calculate the losses in atmospheric medium. The governing equation is defined as

$$\tau = \frac{I_R}{I_O} = e^{-\gamma x} \tag{4}$$

I_R is the intensity at distance x, I_o is the initial intensity and γ is the attenuation coefficient. Attenuation coefficient depends upon atmospheric scattering and absorption coefficients. In addition, can be defined as

$$\gamma = \alpha_m + \alpha_a + \beta_m + \beta_a \tag{5}$$

α and β are coefficients for absorption and scattering by molecular and aerosol particles.

Scattering. It can be considered as the redistribution of energy among the particles of the light signal. This results in the distortion in the waveform of the signal rather than reduction in amplitude. The degree of scattering is proportional to the size of the particle. For smaller sizes $r \ll \lambda/2\pi$, Rayleigh scattering formula is used to calculate the scattering defined as

$$\alpha_s = \frac{fe^4\lambda_o^4}{6\pi\varepsilon_o^2 m^2 c^2 \lambda^4} \tag{6}$$

For $r \approx \lambda/2\pi$, Mie scattering is used and defined as

$$\gamma = \frac{3.91}{V}\left(\frac{\lambda}{550}\right)^2 \tag{7}$$

For $r \gg \lambda/2\pi$, Ray-tracing can be used to study the scattering effect.

Absorption. This is actual loss in signal strength due to the energy absorbed by the medium. Similar to scattering absorption also depends on the size of the particles (σ) and the concentration of the absorbing particles (N). Typically governed by the equation

$$\alpha = \sigma N \tag{8}$$

Further, absorption can be two factors:

a. *Aerosol Absorption.* This absorption is due to larger particles such as particles of dust, water vapor, fog, etc. These are mostly a local phenomenon and varies with temperature, time of day and location.
b. *Molecular Absorption.* This happens when the particles of light intermingle with the molecules present in the atmosphere. These molecules could compose of the various gasses present in the air. When the input light signal strikes, the molecules in the air the molecules starts vibrating and in case this vibration leads to resonance a near complete blocking of the incident light occurs. This is known as blocking-window for a particular spectrum of light.

4 Progress in Quantum Satellite Communication

For worldwide secure communication, it is constitutive to use satellite, timeline of the major milestones and major missions are shown in Fig. 8. First QKD with satellite was done from Los Alamos by the team of Richard Hughes [8]. Simultaneously, demonstration of QKD for high altitude is done in Europe in the presence of the Defense Evaluation and Research Agency of UK and LMU Munich [7]. Additionally, the European Space Agency (ESA) implements quantum communication in space [9, 10].

Some of the prominent small satellite mission details are briefed in the sections below.

CubeSat Quantum Communication (CQuCoM). It was proposed first in the year 2015 [29] by a consortium of 6 nations lead by joint efforts of United Kingdom and Singapore [23, 30– 32]. Initial mission comprises of two launches, the prime objective was to demonstrate the possibility of using nanosatellites as trusted nodes for QKD. Multiple ground station were used including MLRO (Italy). This mission also crystalized the idea that small satellites can be used effectively for establishing QKD at a fractional cost and time in comparison with a standard satellite launch. BB84 protocol with decoy-sates was used to establish QKD for the first mission and for second mission will implement QKD using entangled photons.

Quantum Research CubeSat (QUARC). Launched as a stepping stone to develop a QKD network across UK. This mission helped in developing new technologies for miniaturizing the reliable tracking and pointing devices. One of the primary objective of the mission was to provide a secure communication between UK's critical infrastructure [33]. For this mission, a constellation of 15 CubeSat and 43 optical ground stations (OGS) are setup. These OGS are distributed uniformly across UK [34]. Ideally, the OGS shall be near to the end user (primarily urban centers) but there is a trade-off. If the OGS are placed near to the urban population, the key

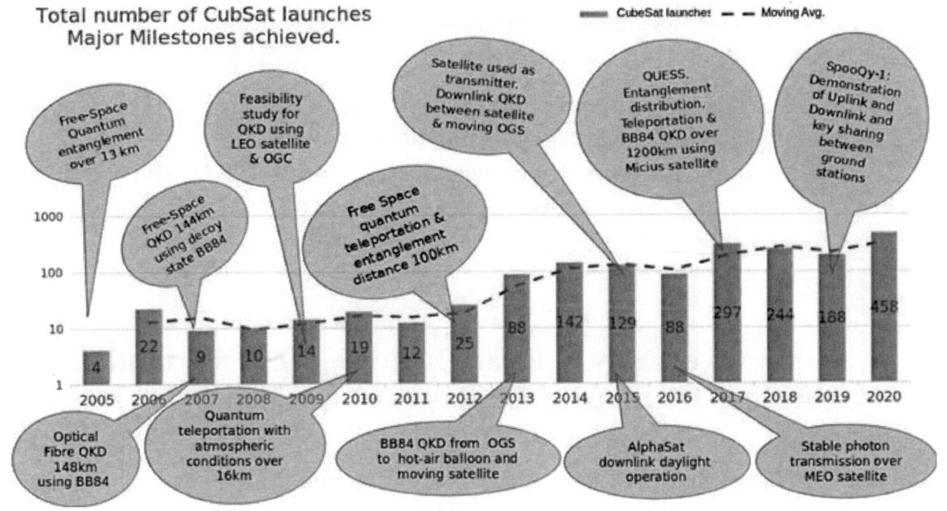

Fig. 8 Milestones and number of CubeSat missions: (a) Free-Space Quantum entanglement over 13 km [17]. (b) Free-Space QKD 144 km using decoy state BB84 [14]. (c) Feasibility study for QKD using LEO satellite & OGC [18]. (d) Free space quantum teleportation and entanglement distance 100 km [19]. (e) Satellite used as transmitter, downlink QKD between satellite & moving OGS [20]. (f) QUESS Entanglement distribution, Teleportation & BB84 QKD over 1200km using Micius satellite [21, 22]. (g) SpooQy -1: Demonstration of Uplink and Downlink and key sharing between ground stations [23]. (h) Optical fibre QKD 148 km using BB84. (i) Quantum teleportation with atmospheric conditions over 16 km [24]. (j) BB84 QKD from OGS to Hot air balloon and moving satellite [25]. (k) AlphaSat downlink daylight operation [26]. (l) Stable photon transmission over MEO satellite [27, 28]

rate decreases due to urban lights and high level of aerosol pollution. A smart way-out is considered by using mobile stations with telescope capable to processing the quantum signal. The mobility helps the OGS to mitigate the issues of the frequent cloud covers and other atmospheric disturbances.

UK National Quantum Technology (UKQT) hub mission. UK-based mission is planning to develop CubeSat constellation in 2023–24 [35]. This will demonstrate space to ground downlink communication. Program will use the learnings and technologies developed in the previous missions SpeQtre and QUARC.

QUBE. It is a German initiative to check the feasibility of QKD between the CubeSat and the OGS situated on ground in Germany [36]. It will use an attenuated pulse to establish QKD.

Q3Sat. Is a 3U quantum CubeSat being developed by Austria using an Uplink configuration [37]. This can be considered as low cost test setup for technology adoption.

NanoBob. An initiative by French and Austrian organizations to build 12U CubeSat [27]. It uses a technology developed by Blue Canyon Technologies to precisely align itself to communicate with an OGS and other quantum satellites. It was found that a sun synchronous orbit shall be used to maximize the key rates but a circular orbit is a much cheaper option and cost effective.

Quantum Encryption and Science (QEYSSat). It a Canadian initiative to develop a low earth orbit micro-satellite weighting around 100 kg in 2022–2023 timeline [38]. The objective is to demonstrate a weak-coherent pulse QKD in an uplink setup. Where in the sender generates a weak-coherent pulse on ground and send it across to the micro-satellite [39].

5 Comparative Study of Basic Quantum Key Distribution Algorithms for Satellite Deployment

Most commonly used protocols for quantum satellite communication are BB84, SARG04, E91, BBM92 and TF-QKD. There are many factors which contribute to the limitations of each and every individual protocol. Hence, it would be only justified if we compare the protocols, which can be used interchangeably or with minor changes.

We have selected BB84 with decoy states, BBM92 and TF-QKD protocols as they all can utilize similar configuration of satellite. All three of them can use polarization as quantum state. Some of the important parameter for the selection is listed in Table 1.

For comparison, we have taken Rx (diameter of receiver telescope) constant at 0.8 m and run the simulation with varying the values of Tx (diameter of transmitter telescope), i.e., 0.08, 0.135, 0.35 and 0.8 m as shown in Fig. 9. Comparison was drawn by plotting the generated secure key rate (bits/sec) versus the elevation angle for the satellite for BB84 with decoy state and BBM92 protocols.

For comparing TF-QKD, we plotted various combinations for Tx (0.8, 0.35, 0.135 and 0.08 m) and FP (0.2, 0.4, 0.6 and 0.8 m) values for a constant Rx of 0.8 m as shown in Fig. 10. This was done to find the maximum secure key rate possible with the TK-QKD protocol (Table 2).

The key generation rate depends upon the size of the telescope used and the size of receiver used. For simplification of comparison, we tested only for one receiver

Table 1 Parameters for selecting suitable QKD algorithm

Name of the protocol	Available quantum states or encoding	Required photon source	Required trusted nodes
Decoy state BB84	Polarization, phase, time bin, OAM	SPDC Weak laser	Yes
BBM92	Polarization	SPDC Weak laser	No
TF-QKD	Polarization	Weak laser	No

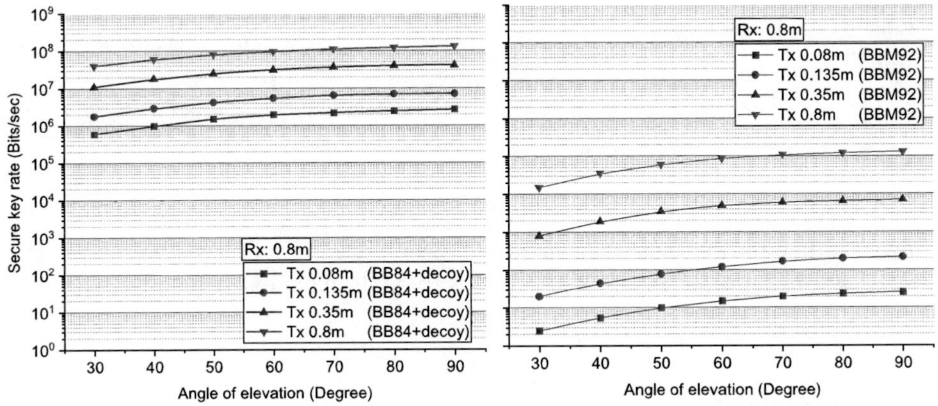

Fig. 9 Graph between "key rate versus elevation angles" **a** BB84 + Decoy state, **b** BBM92

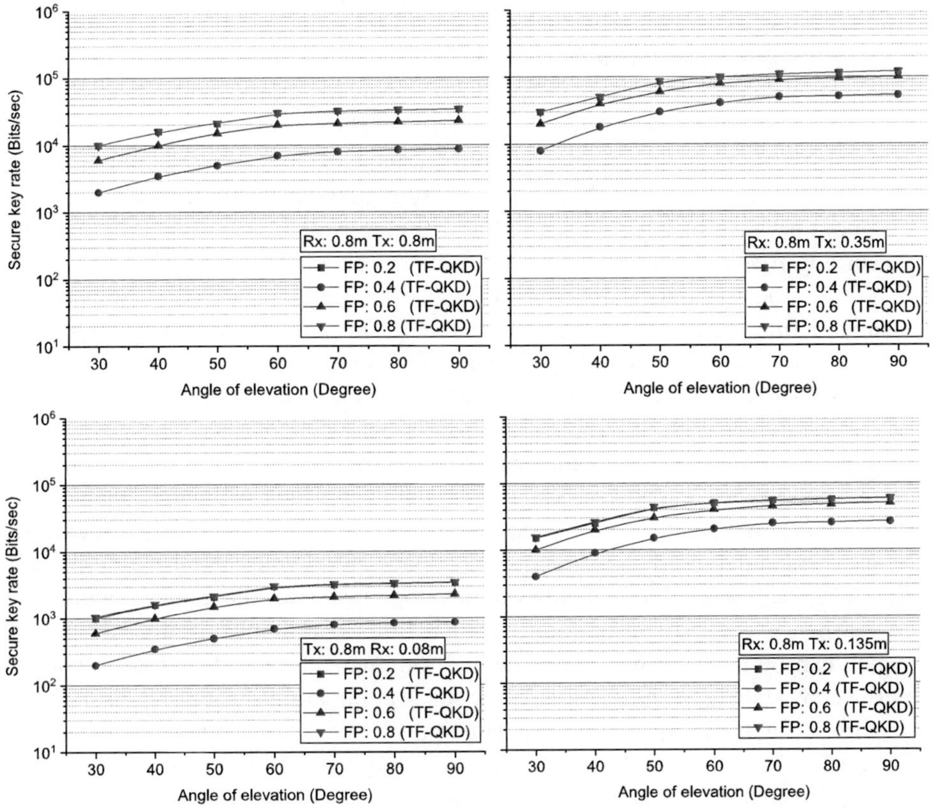

Fig. 10 Graph between "key rate versus elevation angles" TF-QKD

Table 2 Secure key rates generated

Name of the protocol	Secure key rate (Mbits/s)
Decoy state BB84	1–10
BBM92	0.0001–0.001
TF-QKD	0.001–0.01

size. Further, for the same receiver size, BB84 + Decoy state generates the best key rates. Key generations rate of BB84 QKD protocols in our experiment was 10 Mbit/s.

6 Conclusion

Based on the above study, it can be concluded that satellite will play a central role in any formation of global quantum network. It is highly impossible to plan any large area network without quantum satellites. Most commonly used quantum particle use quantum communication are photons and to transfer photons from sender to receiver we need to use a medium or channel. The length of the channel depends upon physical characteristics and design parameters. It can also be concluded that optical links (most commonly used in classical networks) are of little use for large distances because of inherent scattering and non-availability of quantum repeaters. According rough estimates, we would to have a repeater at every 15 km of optical length in a nosy channel. To avoid use of optical fibers, engineers can use optical beams or free-space communication channel. These channels also have their own limitations which are studied in brief in this paper. Still in comparison with fiber links, free-space links provided a multifold increase in the communication length.

In recent year with the improvement in optical technologies, engineers are able to miniaturize the satellites which can be used for quantum communication. These small satellites are called CubeSat. These CubeSat are small and efficient and can be produced at a fractional cost and time of a conventional satellite. CubeSat's can be easily manufactured and launched into space along with the other satellites. This unique advantage enables scientists to develop clusters of these small satellites.

Further, a comparison is done between three commonly used QKD protocols which can be used for quantum satellite communication. It was found that BB84 with decoy state is the most suitable for space applications as it provides the highest secure key generation rates in our simulation when compared with BBM92 and TF-QKD. Going forward, it would be interesting to extend the comparison to other protocols such as coherent one way protocol and SARG04 protocols.

References

1. Shor PW (1994) Algorithms for quantum computation discrete logarithms and factoring. In: Proceedings 35th annual symposium on foundations of computer science. IEEE, pp 124–134
2. Grover LK (1996) A fast quantum mechanical algorithm for database search. In: Proceedings of the twenty-eighth annual ACM symposium on theory of computing, pp 212–219
3. Bennett CH, Brassard G, Breidbart S, Wiesner S (1983) Quantum cryptography, or unforgeable subway tokens. In: Advances in cryptology. Springer, pp 267–275
4. Bennett CH, Brassard G. Quantum cryptography: public key distribution and coin tossing. arXiv preprint arXiv:2003.06557
5. Bennett CH, Brassard G, Breidbart S (2014) Quantum cryptography II: how to re-use a one-time pad safely even if p= np. Nat Comput 13:453–458
6. Lenstra A, Hughes JP, Augier M, Bos JW, Kleinjung T, Wachter C (2012) Ron was wrong, whit is right. Technical report. IACR
7. Kurtsiefer C et al (2002) A step towards global key distribution. Nature 419(6906):450
8. Hughes RJ et al (1999) Quantum cryptography for secure free-space communications. Proc SPIE 3615:98–103
9. Rarity JG et al (2002) Ground to satellite secure key exchange using quantum cryptography. New J Phys 4:82
10. Aspelmeyer M et al (2003) Long-distance quantum communication with entangled photons using satellites. IEEE 9(6):1541–1551
11. Perdigues J et al (2007) Entanglement-based quantum communication over 144 km. Nature 3(7):481–486
12. Ursin R et al (2009) Space-QUEST, experiments with quantum entanglement in space. Europhys News 40(3):26–29
13. Joshi SK et al (2018) Space QUEST mission proposal: experimentally testing decoherence due to gravity. New J Phys 20(6):063016
14. Schmitt-Manderbach T et al (2007) Experimental demonstration of free-space decoy-state quantum key distribution over 144 km. Phys Rev Lett 98(1):010504
15. Pirandola S et al (2017) Fundamental limits of repeaterless quantum communications. Nat Commun 8:15043
16. Pirandola S (2021) Limits and security of free-space quantum communications. Phys Rev Res 3:013279
17. Peng CZ et al (2005) Experimental free-space distribution of entangled photon pairs over 13 km: towards satellite-based global quantum communication. Phys Rev Lett 94(15):150501
18. Armengol JMP et al (2008) Quantum communications at ESA: towards a space experiment on the ISS. Acta Astronaut 63(1):165–178
19. Yin J et al (2012) Quantum teleportation and entanglement distribution over 100-kilometre free-space channels. Nature 488(7410):185–188
20. Bourgoin JP et al (2015) Free-space quantum key distribution to a moving receiver. Opt Express 23(26):33437–33447
21. Yin J et al (2017) Satellite-based entanglement distribution over 1200 kilometers. Science 356(6343):1140–1144
22. Liao SK et al (2017) Satellite-to-ground quantum key distribution. Nature 549(7670):43–47
23. Bedington R et al (2016) Nanosatellite experiments to enable future space-based QKD missions. EPJ Quant Technol 3(1):12
24. Jin XM et al (2010) Experimental free-space quantum teleportation. Nat Photon 4(6):376–381
25. Nauerth S et al (2013) Air-to-ground quantum communication. Nat Photon 7(5):382–386
26. Elser D et al (2016) Quantum measurements of signals from the Alphasat TDP1 laser communication terminal. In: International conference on space optics—ICSO 2016, vol 10562. International Society for Optics and Photonics, p 105623O
27. Kerstel E et al (2018) Nanobob: a CubeSat mission concept for quantum communication experiments in an uplink configuration. EPJ Quant Technol 5(1):6

28. Dequal D et al (2016) Experimental single photon exchange along a space link of 7000 km. Phys Rev 93(1):1–5
29. Lätt S et al (2014) ESTCube-1 nanosatellite for electric solar wind sail in-orbit technology demonstration. Proc Est Acad Sci 63(2):200–209
30. Tang Z et al (2016) Generation and analysis of correlated pairs of photons aboard a nanosatellite. Phys Rev Appl 5:054022
31. Morong W, Ling A, Oi D (2012) Quantum optics for space platforms. Opt Photon News 23(10):42–49
32. Bedington R et al (2013) Deploying quantum light sources on nanosatellites II: lessons and perspectives on CubeSat spacecraft. In: Electro-optical and infrared systems: technology and applications XII; and quantum information science and technology, vol 9648. International Society for Optics and Photonics, p 964811
33. Polnik M et al (2020) Scheduling of space to ground quantum key distribution. EPJ Quant Technol 7(1):3
34. Mazzarella L et al (2020) QUARC: quantum research cubesat—a constellation for quantum communication. Cryptography 4(1):7
35. Quantum Communications Hub: Satellite Quantum Key Distribution for Space (2020) QCH-Satellite-Quantum-Key-Dist-for-Space.pdf (quantumcommshub.net)
36. Haber R et al (2018) Qube—a CubeSat for quantum key distribution experiments. In: 32th annual AAIA/USU conference on small satellites (SSC18-III-05)
37. Neumann SP et al (2018) Q3 Sat: quantum communications uplink to a 3U CubeSat—feasibility & design. EPJ Quant Technol 5(1):4
38. Canadian Space Agency (2020) Quantum encryption and science satellite (QEYSSat). https://asc-csa.gc.ca/eng/satellites/qeyssat.asp
39. Jennewein T et al (2014) QEYSSAT: a mission proposal for a quantum receiver in space. Proc SPIE 8997